The Marriages and Families Activities Workbook

Ron J. Hammond
Utah Valley State College

Barbara Bearnson
Utah Valley State College

THOMSON
WADSWORTH

Australia • Canada • Mexico • Singapore • Spain • United Kingdom • United States

For more information about our products, contact us at:
Thomson Learning Academic Resource Center
1-800-423-0563

For permission to use material from this text, contact us by:
Phone: 1-800-730-2214
Fax: 1-800-730-2215
Web: www.thomsonrights.com

Asia
Thomson Learning
5 Shenton Way #01-01
UIC Building
Singapore 068808

Australia
Nelson Thomson Learning
102 Dodds Street
South Street
South Melbourne, Victoria 3205
Australia

Canada
Nelson Thomson Learning
1120 Birchmount Road
Toronto, Ontario M1K 5G4
Canada

Europe/Middle East/South Africa
Thomson Learning
High Holborn House
50-51 Bedford Row
London WC1R 4LR
United Kingdom

Latin America
Thomson Learning
Seneca, 53
Colonia Polanco
11560 Mexico D.F.
Mexico

Spain
Paraninfo Thomson Learning
Calle/Magallanes, 25
28015 Madrid, Spain

Contents

To the Student

What are your risks of divorce? Do you have healthy dating practices? What type of love do you experience? Where did you learn about sex and how will you someday teach your own children about it? What are your concerns about getting married? What is your cultural and ancestral heritage? How do you feel about death and dying? The answers to these and many more questions are found in this workbook. As you use it, carefully consider each of the assessments and quizzes. Keep in mind that numerous scientific studies were consulted for the assessment development. Our sole purpose as authors and professors was to develop a book that will enable students to learn more about themselves and their family experience. We fully hope that by the time you have worked through this book, you will know far more about yourself and your preparation for family and marriage experience.

We also hope to facilitate a better understanding between you and your partner and other family members. Your partner may use another copy of this book in order to better understand his or her experience. By discussing your findings, feelings, and concerns, you may well improve your understanding of others and their understanding of you. Keep in mind that these assessments are not diagnostic instruments; they are self-awareness instruments based on known scientific findings. Many have been beneficial to students for years. Learning can be enjoyable. As far as family success and experience are concerned, learning about ourselves and how we can improve can be both enjoyable and critically important.

To the Instructor

This workbook is designed to help your students come to better understand themselves and their family roles. It is built on the findings from numerous studies in Sociology, Psychology, Child Development, and Marriage and Family Science disciplines. The basic assumption of the authors is this—the more students know about themselves, their values, their histories, and their expectations, the better prepared and situated they will be for their own family experiences.

This workbook is designed: (1) to facilitate self-awareness in students; (2) to facilitate classroom discussions and teaching by bringing well researched principles into the students' cognitive and emotional learning experience; (3) to inform students about applicable research findings that can be very helpful in their personal life, and (4) to facilitate conversations between students and their partners or other family members.

This workbook is *not* a bibliography of family related studies. Such books are available on the market and certainly fulfill an important function. This workbook is for the students and their needs. Instructors will recognize research-based findings in each of these assessments. None of these assessments and quizzes is a diagnostic instrument nor have they been standardized in any group or population. Each is copyrighted with all rights reserved.

Recommended Uses

(1) Assign to students.

(2) Assign to students and allow students to discuss them in small groups.

(3) Assign to students then discuss with the entire class.

(4) Assign as a library research guide and allow students to find current or pertinent research studies relating to them.

(5) Assign students to discuss these assessments with a partner or family member.

(6) Use as an example of how social science research might be applied to real world issues.

(7) Use as a way to make issues more relevant and personalized to the student.

Acknowledgments

The authors would like to thank Rebecca McClellan and Elizabeth Warner for their extensive work on the preparation of this book. We also express our gratitude to many of our students who have used these assessments and assignments and given us valuable feedback. We also express our appreciation to our families for their support and for the meaningful experiences each day brings to us all.

About the Authors

Ron J. Hammond, Ph.D. is a Sociology professor with over 12 years of college teaching experience. He majored in Family Studies and has taught Marriage and Family Preparation, Sociology of the Family, and other related courses at Brigham Young University, Case Western Reserve University, and at Utah Valley State College. He currently produces educational documentaries, scientific research, and various academic books and workbooks. He was selected as Teacher of the Year at Utah Valley State College, in November 1995; was awarded the NISOD Teaching in Excellence Award, in May 1996; and has been listed in Who's Who among America's Teachers, twice. Ron and his wife, Alisa, are raising their children: Daniel, Krystal, Benjamin, Jacob, Isaac, and Abraham.

Barbara Bearnson, M.S. is a professor with over 22 years of college teaching experience. She has taught in Child Development, Life Span and Human Development, Interpersonal Communications, Family Science, and most recently Sociology of the Family. She has training in numerous Marriage and Family Therapy courses as she works toward her Ph.D. She is also a certified Early Childhood Educator. She has various professional and trade publications including scientific articles to her credit. She has taught at Auburn University and Utah Valley State College. She has been nominated as Teacher of the Year at Utah Valley State College. She currently serves as Chair of the Governing Board for Mountainlands Headstart. Barbara and her husband, LeRoy, have raised four children: Bill, Steve, Ruth, and Robert.

STOP!

Before taking any of the quizzes,
assessments, and scales
in the workbook,
keep in mind that these are
NOT
diagnostic instruments!
They are self-awareness tools, which may
help you to glean information collected in
numerous national studies
and
apply that information to your own
personal life experience.

None of these can be used to predict or diagnose
but
all of them can enhance your understanding of
your personal strengths and weaknesses
so that
you can make informed choices about
your family
and relationships.

Understanding the Genogram

The purpose of the genogram is to allow us to visually observe the structure of our family, its history, and the nature of relationships within it. Sarah's genogram provides an example for you. Sarah was born in 1979 and she has one younger brother. Her mother, Reta, had a miscarriage, followed by cancer and death in 1983. Her father, Robert, never remarried. Robert's parents have been divorced since 1971, and he, Sarah, and John moved in with his mother, Mary. Bill, Robert's father, is a regular visitor to their home but lives alone in a retirement condo. Sarah's grandparents on Reta's side live across the country. Sarah feels close to Martha but not to Sam. Martha has made extra efforts to maintain her relationship with both Sarah and John. Grandpa Sam has talked to Sarah only once since her mother's funeral and that was when he happened to answer the phone. Sarah attends community college and lives at home. She has a steady boyfriend but has no plans to make any long-term commitments. You will notice that we did not include Sarah's aunts, uncles, or other relatives. Genograms can be as detailed or simple, comprehensive or brief as you want them to be. Use the space on the next page to plot your own genogram. Be as comprehensive as you like. Start with yourself and work outward through you family. It is advisable to use a pencil so that corrections can be made.

Sarah's Genogram

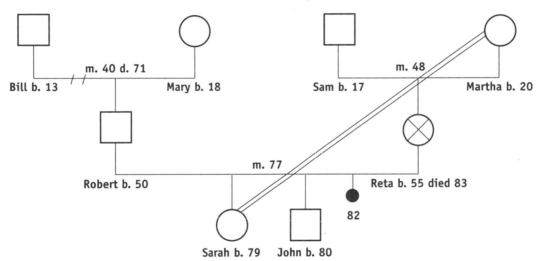

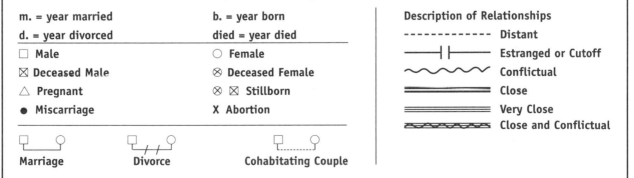

Key to Genogram Symbols

		Description of Relationships
m. = year married	b. = year born	
d. = year divorced	died = year died	
☐ Male	○ Female	
⊠ Deceased Male	⊗ Deceased Female	
△ Pregnant	⊗ ⊠ Stillborn	
● Miscarriage	X Abortion	

- - - - - - - - - - Distant
——┤├—— Estranged or Cutoff
∿∿∿∿ Conflictual
═══════ Close
≡≡≡≡≡ Very Close
⩘⩘⩘⩘ Close and Conflictual

Marriage   **Divorce**   **Cohabiting Couple**

Gender Talk Patterns

This assessment will help you compare your talking patterns to some of the gender communication ideas presented in contemporary self-help books. Think about your own communication patterns as you consider each item. Listening, speaking, and nonverbal communication may vary a certain degree with different people in your relationships. Nevertheless, patterns can be identified for you by using this assessment. Please answer T (True) or F (False) on each of the items below. Consult the key after having finished the entire assessment.

1. T/F   I keep my feelings locked up inside.
2. T/F   I rarely speak in a way that makes me vulnerable to a put down.
3. T/F   I see my relationships in the context of competition not connection.
4. T/F   It is a "dog eat dog" world out there.
5. T/F   You've got to work hard not to be put down.
6. T/F   I hate asking for directions when I'm lost.
7. T/F   I know where I belong in the overall ranking at work.
8. T/F   I don't really know what to say in a group.
9. T/F   I like to know the news so that I can sound informed about things.
10. T/F   I need serious down time to recoup each day.
11. T/F   I am like a warrior of sorts in my everyday life.
12. T/F   Women talk dramatically different from men.
13. T/F   I use my hands when I talk for making gestures.
14. T/F   Men talk differently because their brains work differently from the "women brain."
15. T/F   I like to offer advice so I can fix things for others.
16. T/F   I like to talk a lot.
17. T/F   I rarely speak in a way that makes me stand out among my friends.
18. T/F   I see my relationships in the context of connection not competition.
19. T/F   It's a world of relationships out there.
20. T/F   You've got to work hard not to be socially isolated or disconnected from friends.
21. T/F   It's easy for me to ask for directions.
22. T/F   I speak in a way that makes me equal to others (not superior) in my relationships.
23. T/F   I keep track of my relationships to ensure that they are going well.
24. T/F   I like to hear another's problems and offer similar stories that happened to me.
25. T/F   It's cool when you know someone well enough that they know what you are thinking.
26. T/F   In general, women keep relationships going.
27. T/F   Men talk dramatically different from women.
28. T/F   You have to constantly be on guard because men touch you so much.
29. T/F   I can know how a man feels by simply looking at his facial expressions and body position.
30. T/F   I like to offer solutions that worked for me or my friend; it shows that I care.

4

Key

Questions 1–15 are True (T). These 15 questions represent the typical male patterns of communicating (at least as portrayed in current self-help literature). Questions 16–30 are True (T). These next 15 questions represent the typical female patterns of communicating. How many true answers did you have in each of these sections: 1–15 ___ 16–30 ___? If you marked true in both of these sections then you are probably more like the average person—that is using a variety of communication patterns that overlap the stereotypes of "men talk and women talk." The underlying value of self-help books on how men and women talk differently is not that all men talk one way and all women another. Rather, the value is found in learning patterns of how people talk in general, and learning to see those patterns in the men and women with whom you communicate. They also help us to see that our significant others may simply be different in their communication, as opposed to mean, stubborn, or unsupportive. How does this apply in your relationships? Have someone you interact with regularly take the Partner, Family Member, Friend Version of this assessment, then talk about it after you both have finished.

Gender Talk Patterns

Partner, Family Member, Friend Version

This assessment will help you compare your talking patterns to some of the gender communication ideas presented in contemporary self-help books. Think about your own communication patterns as you consider each item. Listening, speaking, and nonverbal communication may vary a certain degree with different people in your relationships. Nevertheless, patterns can be identified for you by using this assessment. Please answer T (True) or F (False) on each of the items below. Consult the key after having finished the entire assessment.

1. **T/F** I keep my feelings locked up inside.
2. **T/F** I rarely speak in a way that makes me vulnerable to a put down.
3. **T/F** I see my relationships in the context of competition not connection.
4. **T/F** It is a "dog eat dog" world out there.
5. **T/F** You've got to work hard not to be put down.
6. **T/F** I hate asking for directions when I'm lost.
7. **T/F** I know where I belong in the overall ranking at work.
8. **T/F** I don't really know what to say in a group.
9. **T/F** I like to know the news so that I can sound informed about things.
10. **T/F** I need serious down time to recoup each day.
11. **T/F** I am like a warrior of sorts in my everyday life.
12. **T/F** Women talk dramatically different from men.
13. **T/F** I use my hands when I talk for making gestures.
14. **T/F** Men talk differently because their brains work differently from the "women brain."
15. **T/F** I like to offer advice so I can fix things for others.
16. **T/F** I like to talk a lot.
17. **T/F** I rarely speak in a way that makes me stand out among my friends.
18. **T/F** I see my relationships in the context of connection not competition.
19. **T/F** It's a world of relationships out there.
20. **T/F** You've got to work hard not to be socially isolated or disconnected from friends.
21. **T/F** It's easy for me to ask for directions.
22. **T/F** I speak in a way that makes me equal to others (not superior) in my relationships.
23. **T/F** I keep track of my relationships to ensure that they are going well.
24. **T/F** I like to hear another's problems and offer similar stories that happened to me.
25. **T/F** It's cool when you know someone well enough that they know what you are thinking.
26. **T/F** In general, women keep relationships going.
27. **T/F** Men talk dramatically different from women.
28. **T/F** You have to constantly be on guard because men touch you so much.
29. **T/F** I can know how a man feels by simply looking at his facial expressions and body position.
30. **T/F** I like to offer solutions that worked for me or my friend, it shows that I care.

Key

Questions 1–15 are True (T). These 15 questions represent the typical male patterns of communicating (at least as portrayed in current self-help literature). Questions 16–30 are True (T). These next 15 questions represent the typical female patterns of communicating. How many true answers did you have in each of these sections: 1–15___16–30___? If you marked true in both of these sections then you are probably more like the average person—that is using a variety of communication patterns that overlap the stereotypes of "men talk and women talk." The underlying value of self-help books on how men and women talk differently is not that all men talk one way and all women another. Rather, the value is found in learning patterns of how people talk in general, and learning to see those patterns in the men and women with whom you communicate. They also help us to see that our significant others may simply be different in their communication, as opposed to mean, stubborn, or unsupportive. Have someone you interact with regularly take the Gender Talk Patterns version of this assessment, then talk about it after you both have finished.

Gender/Androgyny Role Attitude Assessment

Please answer T (True) or F (False) on each of the items below. If you are married or otherwise committed, then have your partner take the assessment. Compare and discuss only after each has completed it. If you are single, have your parents or close friend take the assessment and discuss it.

1. T/F   Women with school- or preschool-aged children should stay home if at all possible.
2. T/F   Cleaning dishes, laundry, cooking, and so on, are really the woman's responsibility.
3. T/F   Men should be the only breadwinners in the home.
4. T/F   Women are less capable of making important decisions than are men.
5. T/F   Women are naturally dependent on men.
6. T/F   When a woman pursues a career, it's because she has problems with relationships.
7. T/F   When a woman flatters a man to get what she wants, it's O.K.
8. T/F   It would be difficult for me to work for a woman.
9. T/F   You can tell a great deal about a woman by her appearance and sex appeal.
10. T/F   Most women admire the qualities of men and would like to be more like them.
11. T/F   Husbands should really make all the tough decisions in the home.
12. T/F   Women are not as dependable in terms of job stability and commitment.
13. T/F   Women should pursue an education that would directly benefit their homemaking role.
14. T/F   Women are simply not as rational/logical as men.
15. T/F   Women are more social than men.
16. T/F   If she were qualified, I'd vote for a woman for president of the United States.
17. T/F   Lawmakers should support gender equality issues in the legislation they pass.
18. T/F   Women are no more emotional than men tend to be.
19. T/F   Careers provide women with opportunities for self-fulfillment and growth.
20. T/F   Sexuality is enjoyed just as much by women as men.
21. T/F   Men are as capable of loving children as much as women.
22. T/F   Overall, genetics have little to do with the way men and women behave.
23. T/F   Men and women are equally as capable of dominance in society.
24. T/F   Pay should be based on performance, not gender.
25. T/F   Men tend to welcome their wife's earnings in today's tough market.
26. T/F   Neither men nor women are superior to one another.
27. T/F   Both fathers and mothers are essential to the child's upbringing.
28. T/F   The way men and women communicate depends more on their individuality than on gender.
29. T/F   Couples should negotiate housework, yard work, and child-care duties.
30. T/F   The birth of a child is cause enough to celebrate, not its gender.

Scoring Your Gender Role Attitudes

Give yourself **1 point for each True answer in questions 1–15** _____

Give yourself **1 point for each False answer in questions 16–30** _____

Total Score = _____

The closer your score is to 30 points, the more traditional your attitudes tend to be.

Couples and family members enhance the quality of their relationships as they sit down and discuss their gender values and negotiate on those issues that are most significant to those involved. Do these findings accurately reflect you, your expectations, and life experiences? Why or why not?

Personal Men's Issues Assessment

A̲nswer yes or no to all of the events below that have actually occurred or are currently occurring to you. Be as accurate as possible. Females can interview a close male friend or family member. Do not read the answer key until you have taken the assessment.

Yes	Events	No
_____	I was labeled a trouble-maker in public school.	_____
_____	I had learning challenges in public school.	_____
_____	I was not athletic in public school.	_____
_____	I never had a good male role model in my early years.	_____
_____	I feel that I never measure up.	_____
_____	I feel like a mechanical cog in the big economic machinery.	_____
_____	I suffer because of being a male in today's workplace.	_____
_____	My father is not proud of me.	_____
_____	My father is out of touch with issues in my life.	_____
_____	I feel an emotional gap between my father and myself.	_____
_____	I long to be more intimate with my partner.	_____
_____	I was not adequately socialized to nurture children.	_____
_____	I feel emotionally detached from most people.	_____
_____	I do things at work that I regret, but must do to keep the job.	_____
_____	I've thought about suicide before.	_____
_____	I lost custody of my children in a divorce or separation.	_____
_____	I have overheard male bashing comments.	_____
_____	I have been the direct victim of male bashing.	_____
_____	I have suffered discrimination as a male.	_____
_____	I fear that I will be accused of sexual crimes.	_____

Add up all the "Yes" answers. This assessment is designed to measure the intensity of hardship males experience in a post-industrial society. **0 = no hardship, 20 = extreme hardship.** Did these findings surprise you? Discuss them with someone close to you. You can learn more about men's issues on the Internet or in the library.

Softening Patriarchal or Matriarchal
Heads of Families

Often a patriarchal father can lead with compassion, love, and concern for each of his family members (including spouse) without being dictatorial. The mother can also lead the family with firmness and love without a power struggle between the spouse and children.

List ways that the patriarchal father can be the head of the home and still show love and compassion without taking away each individual family member's autonomy:

1.

2.

3.

4.

List ways and circumstances that the mother can be the head of the family and lead in love without power struggles:

1.

2.

3.

4.

How does the balance of power change if newlyweds live with their parents after marriage? Explain:

When the wife and husband both work, what type of shift in power occurs between the spouses? Explain:

Television Messages About Gender Roles

Many studies have established the fact that television viewing shapes our attitudes and outlook on life. Most people in the United States are exposed to numerous television messages while watching 3–4 hours of television per day (that's 9.1 years equivalent by age 65). This project is designed to facilitate an understanding of the gender role messages you get from various television shows. Watch two separate shows from 7–11:00 P.M. and use this table to analyze their presentation of male and female roles.

Factors to Consider	Title of First Show	Title of Second Show
Are the central characters male, female, or both?		
Which characters are shown as being in control or having the most power?		
Are the male characters portrayed as being competent in their social roles? How can you tell?		
Are the female characters portrayed as being independent and capable in their roles? How can you tell?		
If the show is set in the context of a central family, are the male and female characters realistically portrayed? If not, why?		
List three words that basically describe the male characters.		
List three words that basically describe the female characters.		
When comparing yourself to the main character(s) of the same sex, what is the difference (if any)? Would you be friends in real life?		
Are any of the male or female characters exploited sexually, financially, or socially? If so, how?		
Is violence used to coerce male or female characters?		
Overall, are male or female characters more successful on the show? Why?		

Planning Your Family Life Cycle

1. First, fill in the age you will probably be in each of the following "turning points" in your family life cycle. If you've already hit a turning point, put date in the appropriate box.

Turning Point	Probable Age
Get married	
Have your first child	
First child starts school	
First child starts junior high	
First child leaves home (college, military, etc. . . .)	
Last child leaves home	
Retire from profession or career	
One member of the couple passes away	
Second member of the couple passes away	

2. Now compute how many years you will probably spend in each of the following stages of your "family life cycle." (From: Duvall, E. & Miller, B. *Marriage and Family Development,* 6th ed. New York: Harper & Row, 1985.)

Stage	Years in Stage
Stage I: Pre-children Married and no children	
Stage II: Childbearing families Birth of first child to birth of last	
Stage III: Preschool No children in school yet	
Stage IV: School-aged children Ages 6–13	
Stage V: Teenagers Ages 13–19	
Stage VI: Launching centers Children leave home for college, marriage, etc. . . .	
Stage VII: Empty nest All children left home, not yet retired	
Stage VIII: Aging family Includes retirement and the golden years	

3. Describe the biggest surprise you had in trying to estimate what your family life cycle will probably be.

4. Describe what you are going to do differently now to prepare yourself for your entire life cycle.

Life Cycle

Divide the circle into stages of time that might occur in your life.

1. Married, Pre-Children Years
2. Child Bearing Years
3. Pre-School Years
4. School Children Years

5. Teen Years
6. Launching Years
7. Empty Nest Years
8. Aging Family Years

Assuming you will live for the next 60 years, plan your life stages in yearly segments and then place them on the pie chart below. For example if you married at age 25 and waited until you were 30 to have your first child then stage 1 above would be 5 years duration. If you had your last child at age 33 then you would have been in stage 2 for 3 years. Use similar reasoning for the remaining stages. Estimate these for each stage above and graph them in the pie chart below (keep in mind that the entire pie equals 60 years). There will probably be some overlap. Shade in those areas lightly. Once the pie chart is completed it will give you a clear visual representation of your family life cycle stages and how many years you plan to spend in each one. *Definitions for each stage are in the previous assignment.*

Nontraditional Life Cycles:
Variations on the Theme

One drawback of the family life cycle approach is its tendency to focus on intact nuclear families. Millions of people today experience some form of variation from intact nuclear families. This assignment is designed to facilitate a better understanding of differing life cycles. There are two basic components. First, interview a friend or family member who did not pursue the intact nuclear life course. Plot his or her course by inserting the dates of major life events in the appropriate boxes. Second, plot your own life cycle events using dates. Estimate future life events based on your goals and ambitions.

Event	Date of Event for Friend or Family Member	Date or Estimated Date for You
Teen pregnancy		
Single parenting		
Adult unwed pregnancy		
Extended singlehood		
Cohabitation		
First marriage		
First divorce		
Single parenting		
Second marriage		
Second divorce		
Single parenting		
Other?		

Rites of Passage

R ites of passage signal to society that an individual has changed his or her role in society. The right of passage is based on what the culture expects of individuals at certain ages. This expectation is affected by family ethnicity and culture. Some of these expectations may include starting grade school, changing from grade school to middle school, the onset of puberty, getting a driver's license, turning eighteen, graduation, weddings, funerals, religious advancement, and others.

Some adolescents may decide their own rites of passage differently than their parents or adults prescribe. They view becoming sexual, smoking, using drugs or alcohol, profanity, and disregarding authority of parents and society as signals they are now adults.

Evaluate yourself based on your culture and ethnic background. What were the rites of passage expected of you? Did you fall into any adolescent rites of passage behaviors during adolescence? Explain.

1.

2.

3.

4.

Your partner:

1.

2.

3.

4.

How could your experience growing up help your children understand their rites of passage but avoid problematic perceived rites of passage? Explain.

Using the SMART Approach to Evaluating Information

In this exercise, find a news article or report about a scientific study. Use the **SMART** approach as a way of analyzing the information. Answer questions in these basic areas: Sample, Methods, Attitude of skepticism, Researcher bias, and Thorough understanding of the literature.

Read a scientific report, study, or news article and answer the questions below.

Sample

What type of sample was drawn? Convenience _____ Random _____

Does the structure of the sample represent the structure of the population it claimed to represent? ___ Yes ___ No

Is it a SLOP poll? _____ Yes _____ No

Does it have obvious selection biases in sample participants? _____ Yes _____ No

Method

Which method was used to gather the data?

survey _____ Yes _____ No   **participant observation** _____ Yes _____ No

nonparticipant observation _____ Yes _____ No   **experiment** _____ Yes _____ No

field experiment _____ Yes _____ No   **clinical** _____ Yes _____ No

Was it: qualitative _____ or quantitative _____?

Do the conclusions drawn fit the research method used? _____ Yes _____ No

Attitude of Skepticism

Does it simply reinforce commonly held beliefs or is it suggesting a new outlook or perspective?
_____ reinforce _____ new outlook?

Do you agree with its claims? Why? Why not

What do you suspect might be different if it were studied more?

Researcher Bias

Which organization, group, party, performed or paid for the study? _____

Are there obvious areas where findings could be self-serving? ___ Yes ___ No

Did the researchers mention these areas or make an effort to scientifically control for them? ___ Yes ___ No

Thorough Understanding of the Literature

Is this a major ground-breaking discovery or claim? ___ Yes ___ No

Does it appear to coincide with other studies in this field of research? ___ Yes ___ No

Overall Assessment

Based on your evaluation of the information and the answers you gave above, using this form, how would you rate its overall quality? **1 = garbage/junk. Up to 10 = good scientific information.**

My evaluation = _____

Your Dispositions Toward Theories

Make sure to respond to all of the statements below before reading the answer key. There are 30 statements below. Read the statement then place either: "**A**" for "*applies to me,*" "**SA**" for "*sometimes applies to me,*" or "**NA**" for "*never applies to me.*" For each statement, think about your own views on people and society. Remember that this assessment in no way attempts to establish right or wrong views. Its purpose is to help you identify your own perspectives and then connect them to the major theories.

Answer: A, SA, or NA	Statements
_____	1. There are always winners and losers.
_____	2. Certain elements of society provide for critical needs of its members.
_____	3. It is important to consider the meanings people place on things.
_____	4. People try to get the most out of their relationships and interactions.
_____	5. To really understand people you have to look at their entire life course.
_____	6. A family is comprised of many complex relationships.
_____	7. The average person is exploited by those who are more powerful.
_____	8. Society tends to find a balance in spite of its troubles.
_____	9. Sometimes two people see the same thing in two different ways.
_____	10. Most people try to avoid pain and losses.
_____	11. We all go through certain stages in life.
_____	12. Sometimes family members cross appropriate lines in inappropriate ways.
_____	13. Those with power usually get their way.
_____	14. Sometimes certain parts of society break down.
_____	15. How we define our life and life events plays an important role in how we experience those events.
_____	16. There's always a give and take in relationships.
_____	17. We tend to grow into the people we now are.
_____	18. If there is trouble in one family relationship, it often spills over onto other family relationships.
_____	19. Conflict is an inevitable part of social interaction.
_____	20. Certain elements of society don't always do exactly what they were designed to do.
_____	21. People fill their roles in life based on what they think they are supposed to do in those roles.
_____	22. People need to feel that they get some level of value from their relationships.
_____	23. Sometimes people don't go through "normal" life stages.
_____	24. There is a balance in families that is disrupted at times.

_____	25.	Those who **don't have** must struggle with those who **have** for resources.
_____	26.	Sometimes one segment of society accidentally does something either useful or harmful to the rest of society.
_____	27.	How we communicate influences the quality of our relationships.
_____	28.	People typically cut off relationships that are too hurtful or costly.
_____	29.	None of us goes through exactly the same life stages in exactly the same way.
_____	30.	Most families are both functional and dysfunctional.

In the boxes below, circle the statement number if you put either "A" or "SA" in your answer for that statement. If you circled more in one theory than in the others you may lean toward that theory in your personal views. Some find that these statements make sense and are somewhat obvious. That may be because they represent the basic underlying principles of these social theories that have been tested and retested over recent decades. In other words, these theories have a close fit to the real social world in which we live. This assessment should help you to connect these theories to your personal life experience. The more you learn about these theories, the clearer this assessment and its findings will become.

1, 7, 13, 19, 25	2, 8, 14, 20, 26	3, 9, 15, 21, 27
Conflict Theory	Functional Theory	Symbolic Interaction Theory
4, 10, 16, 22, 28	5, 11, 17, 23, 29	6, 12, 18, 24, 30
Social Exchange Theory	Developmental Theory	Family Systems Theory

Critical Life Events Log

Throughout the life course, many significant life events occur in our lives. We often forget or try to forget them for various reasons. Yet, even if forgotten, these events may still impact your life. Each column represents a year in your life (year is indicated at the top of the column). In the columns below, stack the letter and shape of the events as indicated. *See example on following page of Gina's life events.* Record your symbols and their meanings in the appropriate log to the right using these guidelines:

Normal Life Events put first letter and ○ circle (i.e.: Started Kindergarten =   , Graduated High School =   , Started Dating =   , etc. . . .)

Traumatic Life Events put first letter and □ square (i.e.: Accident =   , Divorce =   , Death in Family =   , etc. . . .)

Significant Positive Event put first letter and △ triangle (i.e.: Won Sweepstakes =   , Fell in Love =   , Birth of Child =   , etc. . . .)

Most Significant Events in your life and ∞ infinity (i.e.: Met Spouse =   , Moved to Maine =   , etc. . . .) *These are simply examples, make up and use symbols that work for you.* After you log your life events, answer the following questions designed to help you understand their significance in your life's course and experiences.

Questions

Normal Life Events

1. Which years were major turning points in your life?
2. Looking back, which normal life event was the most helpful to you? What might have been different in your life, had that event *not* occurred?
3. Which normal life event *did not* occur in your life that you see would have made a significant difference?

Traumatic Life Events

1. Which were the years of trauma in your life?
2. Of all the traumatic events, which one appears to be most damaging at this point in your life? Why? Can anything be done now to reduce the negative impact on you from this trauma?
3. What would have been the worst traumatic thing that could have happened but *didn't?* Why?

Significant Positive Events

1. In which years did the positives occur?
2. Which was the single most positive significant event in your life? Why? How much did you personally do to bring about this event?
3. Where would you be had the event *not* occurred? Why?

Most Significant Life Events

1. What year or years were the most significant? How did the events change the course of your life?
2. If you could undo any one of them, which would it be? Why?
3. Which significant events lie before you in your life? What will they bring?

	Normal Life Events Log ○	Traumatic Life Events Log ☐	Significant Positive Events Log △	Most Significant Life Events Log ∞

First Twenty Years of Your Life

01	02	03	04	05	06	07	08	09	10	11	12	13	14	15	16	17	18	19	20

Second Twenty Years of Your Life

21	22	23	24	25	26	27	28	29	30	31	32	33	34	35	36	37	38	39	40+

First Twenty Years of Your Life

	01	02	03	04	05	06	07	08	09	10	11	12	13	14	15	16	17	18	19	20
					Ⓢ	Ⓜ			⒟ⓜ			Ⓐ	Ⓐ Ⓐ	⒡ⓡ Ⓐ	Ⓜ Ⓐ			Ⓖ Ⓓ Ⓙ		△

Second Twenty Years of Your Life

	21	22	23	24	25	26	27	28	29	30	31	32	33	34	35	36	37	38	39	40+
					∞ MJ ∞			∞ TB		Ⓑ	Ⓓ Ⓒ ∞ DJ Ⓢ		△ △N		∞ ML					

Normal Life Events Log ○	Traumatic Life Events Log □	Significant Positive Events Log △	Most Significant Life Events Log ∞
Ⓢ Began school	⒟ⓜ Death of mother	△ Began college	∞ Met and MJ married Jeff
Ⓜ Began music lessons	⒡ⓡ Father remarried	△ Graduated college	∞ Twins TB Born
Ⓖ Graduated high school	Ⓜ Moved out of state	△N New job as music teacher	∞ Divorced DJ Jeff
Ⓓ Began to date	Ⓐ Abused by neighbor		
Ⓙ First full-time job	Ⓢ Single parenting began		∞ Met ML Lance
∞ Dropped out of college	Ⓑ Bankruptcy		
Ⓒ Began counseling	Ⓓ Divorced Jeff		

How Others Influence Your Self-Concept

Pick one role in your life that is most important to you (i.e.: wife, husband, lover, mother, father, best friend, employee, etc.). Using the 12 steps below, analyze your performance in this role and discover how other people in your life shape your self-concept in this role.

Step 1. List the role: _____ How long have you been in this role? _____

Step 2. How important is the role in your life on a scale of 1–10? *Circle one*

1—not very important, 2, 3, 4, 5, 6, 7, 8, 9, 10—very important

Step 3. List the names of five others you personally know who are in the same role.

Step 4. Of those listed in step 3, who performs their role best and worst? Rank them all.

Best: Middle: Second Worst:

Second Best: Worst:

Step 5. Compare yourself in this role to the others in the ranked list. Draw a star at the place you feel you rank.

Step 6. What is it that those who rank better than you are doing that gets them a higher evaluation?

What are those worse than you not doing that you do?

Step 7. Reflect upon the ranking and where you placed your star. Does this enhance or detract from your self-concept? Enhance ___ Detract ___

Briefly describe why:

Step 8. How important is it for you to perform as well as others in the same role?

1—not very important, 2, 3, 4, 5, 6, 7, 8, 9, 10—very important

Step 9. Now, grade yourself on how you fulfil this role:

A, B, C, D, or F

Step 10. List the names of the two most important people in your life.

(1)   (2)

Step 11. Mark the score they would give you if they graded you in this role.

(1) A, B, C, D, or F   (2) A, B, C, D, or F

Step 12. Look at your grade in step 9 and the two grades in step 11. Is there a difference? Yes ___ No ___

How important is this to you? 1—not very important, 2, 3, 4, 5, 6, 7, 8, 9, 10—very important

Our self-concept is shaped not only by our own assessment of ourselves but by the assessment of those with whom we interact. Consider doing this analysis for another important role in your life.

How Colonial Is Your Family?

Answer the questions below based on the family you grew up in. Be as accurate as possible. Don't read the answer key until after you have finished the assessment.

Yes	Questions	No
_____	The main theme of my family was making ends meet.	_____
_____	The children began to do chores as early as possible.	_____
_____	Discipline was harsh at times.	_____
_____	The collective family needs were more significant than individual needs.	_____
_____	Our family was the center of our everyday lives.	_____
_____	Our family was deeply connected to another similar family.	_____
_____	Religion was a main theme in our family.	_____
_____	We produced many of our own goods at home.	_____
_____	Father's rule was most important in our family.	_____
_____	You had to learn skills to do our family work.	_____
_____	We learned the basic reading, writing, and arithmetic at home.	_____
_____	We had to get along with each other.	_____
_____	My own feelings were often ignored.	_____
_____	My needs and wants were not important.	_____
_____	We were not very wealthy.	_____
_____	My parents had a big say in whom I dated.	_____
_____	Romance was not as important as hard work in a potential mate.	_____
_____	My parents did not really love each other.	_____
_____	One of my parents was subordinate to the other most of the time.	_____
_____	Childhood was really more like miniature adulthood.	_____
_____	Our parents still have strong control in our lives.	_____

Key

Add up all the "yes" answers. A score of **21** represents very high colonial traits. A score of **0** represents very low traits. Do you have concerns about any of the issues above? Did anything surprise you about your family? So many of our modern day families have traces of the colonial family in them. Perhaps you can discuss these issues with a family member.

Food and Meals in My Home

In this assessment, answer these questions based on the family you grew up in or the family you currently live in (whichever is the most influential).

1. There are typically 21 meals in any given week. On the average, how many of those meals are restaurant meals? ___/21

2. On the average, how many meals per week are precooked, microwaved, or oven heated? ___/21

3. On the average, how many meals per week are made from basic ingredients (scratch)? ___/21

4. Is there a daily or weekly meal that everyone typically attends? If so, describe where, when, who, what, and how it occurs.

5. During one of these meals mentioned in question 4, what is the typical series of events that transpire? (How many courses, do people talk in multiple or singular conversations, does mother, father, siblings serve, who cleans up, and how much socializing occurs throughout the prep, eating, and cleanup stages of the meal?) Write a short descriptive paragraph.

6. If there were one word to describe the family meal experience (questions 4 and 5) in your home, then what would it be?

7. What would you like to see more of at this meal?

8. What would you like to see less of at this meal?

9. Describe the other meals throughout the week in a brief sentence.

10. What family of origin meal traits and patterns do you feel strongly about in terms of repeating them in your own family?

11. What family of origin meal traits and patterns do you want to avoid in your own family?

12. Discuss these questions and your answers with a family member, partner, or friend. What are their observations?

13. In reference to questions 1–3 above, the U.S. pattern for meal consumption and preparation is about ⅓ eating out, ⅓ precooked heat and serve, and ⅓ made from scratch. How does your family compare?

Family Foods and Meal Traditions

In this assessment, you will consider family traditional meals and dishes. This is an assessment designed to help you identify the origin of the foods you typically eat with your family. Reflect upon current and past meal patterns when answering. This assessment might be best filled out while talking to parents, grandparents, and other extended family members.

1. List your 10 favorite family traditional dishes, desserts, or meals.

   1.   5.   9.
   2.   6.   10.
   3.   7.
   4.   8.

2. What is the earliest know origin of these dishes in your family?

   1.   5.   9.
   2.   6.   10.
   3.   7.
   4.   8.

3. Which of the 10 items listed above is the single most favorite in your family?

4. Who among your family members made the same dish?

   Parents ___Yes ___No   Grandparents ___Yes ___No

   Uncles/aunts and/or great uncles/aunts ___Yes ___No   Great grandparents ___Yes ___No

   Great, great grandparents ___Yes ___No   Great, great, great grandparents ___Yes ___No

   Others? ___Yes ___No Who?

5. Are there deep rituals, traditions, or meanings placed upon any of the dishes listed above? Briefly describe it here:

6. Describe an emotionally meaningful experience you personally have had involving one or more of these foods, which contributes to its continued use in your family meals.

7. Is there a dish, dessert, or meal that you no longer prepare or eat?

8. If so, why did it fall from tradition? What would happen if you began to eat it again?

Attitudes About Food in Your Family of Origin

In this assessment think about the meaning of food in your family of origin and the context in which it was prepared and eaten. Answer T (True) or F (False). Wait until the assessment is completed before looking at the key.

1. T/F   Food came with fun in our home.
2. T/F   Almost every time we got together, we ate something.
3. T/F   We were made to feel guilty when eating food.
4. T/F   We had to clean our plates at every meal.
5. T/F   We always overdid it when it came to food.
6. T/F   Food was a necessary evil.
7. T/F   There was really never enough food in our home.
8. T/F   We loved to eat when company came over.
9. T/F   Somehow, food made the feelings more positive in our home.
10. T/F   When things got stressful, we would eat more food.
11. T/F   I feel ashamed for what I eat.
12. T/F   We typically ate balanced meals.
13. T/F   Drinks are an important part of the eating experience.
14. T/F   We all enjoyed certain favorite family foods.
15. T/F   At funerals and other sobering family gatherings, there was food.
16. T/F   Family members had to be skinny to fit in.
17. T/F   Everybody helped to plan the weekly menus.
18. T/F   We often celebrated birthdays with food and drink.
19. T/F   Holidays meant lots of food.
20. T/F   We never ate while watching the television.
21. T/F   The same family members typically prepared and cleaned up the food.
22. T/F   I use the same recipes my parents and grandparents used.
23. T/F   Food helps to get through the tough times.
24. T/F   Whenever we ate, we talked and interacted.
25. T/F   Food was eaten in religious contexts.
26. T/F   We were a calorie-counting family.
27. T/F   We eat some of the same foods we ate decades ago.
28. T/F   Food budgets get top priority in our house.
29. T/F   Food kind of says "I love you."
30. T/F   Food brings all of us together.

Key

1 = T, 2 = T, 3 = F, 4 = F, 5 = F, 6 = F, 7 = F, 8 = T, 9 = T, 10 = F, 11 = F, 12 = T, 13 = T, 14 = T, 15 = T, 16 = F, 17 = T, 18 = T, 19 = T, 20 = F, 21 = T, 22 = T, 23 = F, 24 = T, 25 = T, 26 = T, 27 = T, 28 = T, 29 = T, 30 = T

Your Score _____

This assessment is biased with the assumption that food is useful and meaningful in facilitating positive social interaction among family and friends. A score of **30** indicates healthy attitudes about food in the family context. A score of **0** indicates very unhealthy ones. Do any of the items above stand out as significant concerns for you about food in your family of origin? If so what impact does it have on your current food experience? What can you do to address this and any other concern?

How Metropolitan Are You?

L iving in or around a large metropolitan area is different from living in the rural, less-populated countryside. In the United States today, most people live in suburban or urban settings. The closer to the city you want to live the more metropolitan you tend to be. This assessment indicates how metropolitan you really are. Answer each question either T (True) or F (False) about yourself. Consult the key after having completed the assessment.

1. T/F   I feel safe having many people close by.
2. T/F   I love the diversity of people.
3. T/F   I like to focus on my personal life and not be bothered with the details of the lives of others.
4. T/F   I am very tolerant of differences among people.
5. T/F   Riding the metro is intriguing to me.
6. T/F   I thrive on museums, art exhibits, shows, and so on.
7. T/F   I enjoy many different types of foods.
8. T/F   Big cities are stimulating.
9. T/F   I need access to major sporting events.
10. T/F   I like being a part of/setting the latest trend.
11. T/F   Families need to be relatively small.
12. T/F   I prefer an apartment over a piece of land with a house on it.
13. T/F   I like to visit ethnic sections of the city.
14. T/F   Traffic is part of life.
15. T/F   I enjoy the social scene found in the city.
16. T/F   The city has its problems but that's part of life.
17. T/F   There are strange people in the cities just as there are in the countryside.
18. T/F   I like the freedom to be who I really am.
19. T/F   Many city dwellers are better educated.
20. T/F   I like the sometimes unpredictable nature of city life.
21. T/F   Crime happens everywhere.
22. T/F   I like to see old neighborhoods get remodeled.
23. T/F   I have fond memories of certain buildings, parks, and so forth, in the city.
24. T/F   The city is home.
25. T/F   There are many people like myself in the city.
26. T/F   Sometimes I like to leave the city for a vacation.
27. T/F   Children grow up better in the city.
28. T/F   The best shopping is in the city.
29. T/F   The city kind of grows on you.
30. T/F   It takes sophistication to live in the city.

Key

1–30 = True

Your score _____

This assessment is obviously biased toward cosmopolitan lifestyles and preferences. A score of **30** indicates a very high level and a score of **0** indicates very low levels of this lifestyle. How do your preferences for cosmopolitanism play into your life, life cycle, and life pursuits? Discuss these issues with your spouse, close friend, or family member. If you could live in accordance with these preferences what would have to be traded off (if anything)? Do you and your partner/spouse see eye to eye on these issues?

Singles Well-Being Assessment

In this assessment answer the questions T (True) or F (False) in reference to your current or more recent role as a single. Be as accurate as possible and do not look at the key until this assessment is completed.

1. T/F   I have good relationships with people.
2. T/F   I give and take with my friends.
3. T/F   I feel comfortable in my status as single.
4. T/F   When I feel pressured by family or friends to marry, I deal with it well.
5. T/F   Overall, I have a balance in my life.
6. T/F   My career is going where I want it to go.
7. T/F   I get adequate exercise.
8. T/F   I am of worth just as I am.
9. T/F   My loneliness never overwhelms me.
10. T/F   I have little anxiety about my future.
11. T/F   I have a life plan and follow it.
12. T/F   I enjoy my life.
13. T/F   I am not afraid of taking appropriate risks.
14. T/F   I enjoy my freedom of spontaneity.
15. T/F   I have balance in my sex life.
16. T/F   I am self-sufficient.
17. T/F   I would not change things in my life right now.
18. T/F   I enjoy the children of my friends and family.
19. T/F   I am becoming my full self.
20. T/F   If I wanted to marry then I would.
21. T/F   I often help out family and friends when they need it.
22. T/F   I'd like to make more money.
23. T/F   I feel as though I fit in most of the time.
24. T/F   I often go out.
25. T/F   I get plenty of sleep.
26. T/F   I feel that my emotional intimacy needs are being met.
27. T/F   Society is more accepting of singles today.
28. T/F   I get along well with my friends who have married.
29. T/F   Life really is fun.
30. T/F   My educational needs are met.

Key

1–30 = T

Your score _____

This assessment is biased toward healthy balance in your life as a single. Look at any F (False) answers you made and consider the relevance of this issue to your overall health and well-being as an individual.

Which of these issues above are the most critical to you at this point in your life?

Which of these issues above have little or no relevance in your life at this time?

Which of these issues above need to be dealt with in society at large?

Discuss these findings with a friend or family member. Identify areas of improvement and set some goals.

How Safe Was the Home You Grew Up In?

Answer the following questions T (True) or F (False) as accurately as possible. If you moved and lived in more than one home, answer true if any questions apply. Do not read the key until after you've completed the questions.

1. T/F   Family members were injured in accidents involving the house.
2. T/F   Our doors and windows locked securely.
3. T/F   There was adequate living space for our size of family.
4. T/F   I grew up in a safe neighborhood.
5. T/F   My family was victimized by crime when I lived with them.
6. T/F   We could express our true feelings in our home.
7. T/F   My parents argued excessively.
8. T/F   Sometimes physical violence occurred between family members.
9. T/F   Everyone in our home had to be happy all the time.
10. T/F   If we cried we were reprimanded.
11. T/F   Sexual abuse never took place in our family.
12. T/F   Emotional abuse never took place in our family.
13. T/F   Physical abuse never took place in our family.
14. T/F   My belongings were never stolen by family members.
15. T/F   My belongings were never stolen by friends of the family.
16. T/F   Relatives never hurt me.
17. T/F   Relatives never hurt any of our family members.
18. T/F   Adult children never returned to live in our home after they married.
19. T/F   My needs were always met while I was growing up.
20. T/F   I had to take care of my parents while I was growing up.
21. T/F   Family members knew too much about each other's lives.
22. T/F   People often came to our house to booze it up in parties.
23. T/F   Illegal drugs were never used in my home.
24. T/F   Weapons were never used in my home.
25. T/F   My parents protected the smaller children from older one's who could hurt them.
26. T/F   None of my family members were arrested while I was growing up.
27. T/F   I want the family I make as an adult to be just like the one I grew up in.

Key

This assessment is biased toward unsafe homes. Grade your answers using this key.

1 = T; 2 = F; 3 = F; 4 = F; 5 = T; 6 = F; 7 = T; 8 = T; 9 = T; 10 = T; 11 = F; 12 = F; 13 = F; 14 = F; 15 = F; 16 = F; 17 = F; 18 = F; 19 = F; 20 = T; 21 = T; 22 = T; 23 = F; 24 = F; 25 = F; 26 = F; 27 = *

Add up all the correct questions.

Your Score _____

Maximum score is **26**, suggesting a very *unsafe* family of origin. A score of **0** would suggest a very *safe* family of origin.

*Question 27 is neither true nor false but it is significant because it indicates how and why we make our own adult families, for better or for worse, under the influence of our family of origin, even if our family of origin was not altogether safe. Discuss these findings with a family member or close friend.

Family of Origin Enmeshment

Answer the following questions T (True) or F (False) about your relationship within your family of origin (*you can repeat the assignment for a friend or another close relationship later if desired*). Be as honest and accurate as possible as you reflect upon and answer these questions. Wait until you have completed the assessment to read the key.

1. T/F   We are extremely close.
2. T/F   We have been close for a very long time.
3. T/F   We know everything about each other.
4. T/F   When something happens to one of us we share every detail.
5. T/F   When one of us in a mood, the other will get into the same mood.
6. T/F   We tell secrets to each other that have been confided by others.
7. T/F   We are a strong family.
8. T/F   If one of us struggles, everyone else feels down about it.
9. T/F   Most of the time, family matters are very serious.
10. T/F   If two of us were talking, any other family members could join in at any time.
11. T/F   Keeping secrets is a sign of family weakness.
12. T/F   If two family members disagree, it causes the entire family to feel bad.
13. T/F   Certain family members must be protected because they are fragile.
14. T/F   It is so rude to disagree with parents.
15. T/F   Many of our problems never get worked out.
16. T/F   Each family member was allowed to have his/her own viewpoint.
17. T/F   We never opened each other's mail.
18. T/F   Each of us is very emotionally independent of the others.
19. T/F   Family disagreements were casual events, not such a big deal.
20. T/F   Most family problems eventually worked themselves out.
21. T/F   Most family members were physically healthy most of the time.
22. T/F   Our individual uniqueness was eagerly welcomed.
23. T/F   We had certain issues that simply were not discussed because they were too private.
24. T/F   It was/is easy for most children in our family to move out on their own as adults.
25. T/F   I don't need my parents' approval before I do something significant.
26. T/F   We self-disclose enough but not too much in our family.
27. T/F   We know very little about each other's sex lives in our family.
28. T/F   It's safe to tell secrets in our family.
29. T/F   We can forgive another's mistakes given enough time.
30. T/F   Some of us are closer to the entire family than others.

Key

This assessment is biased toward enmeshed families. The higher your score, the greater the enmeshment in your family of origin.

1–15 = T and 16–30 = F.

Your Score _____

A score of **30** is extremely enmeshed. A score of **0** is extremely balanced. Enmeshment is defined as an overinvolvement, overconnection, and overidentification with family members that results in a lack of individual identification. If this definition does not make sense go to the library or Internet and learn more about it and its consequences. Talk to a close family member or friend about your score and these issues.

Your Religiosity Continuum

This assessment is designed to help you discover your levels of religiosity or participation in formal religion and its meaning in your personal life. First, write your initials on the continuum below in the place that best represents your overall religiosity. Then answer the questions T (True) or F (False) below as accurately as possible.

Very Religious ——————————————— Neutral ——————————————— Not Very Religious

Questions

1. **T/F** I consider myself to be a member of a religious denomination.
2. **T/F** I attend worship services regularly throughout the year.
3. **T/F** I consider myself to be an "active" member.
4. **T/F** I regularly volunteer time to religious activities.
5. **T/F** I regularly donate money to my religion.
6. **T/F** I worship even when I'm not in formal worship services (i.e.: at home, work, etc. . . .).
7. **T/F** I often meditate upon my beliefs.
8. **T/F** I read sacred writings on my own.
9. **T/F** I pray regularly.
10. **T/F** I live my beliefs.
11. **T/F** I believe in a "higher power."
12. **T/F** I need the influence of a "higher power."
13. **T/F** I enjoy the relationships I have with other members of my faith.
14. **T/F** I receive support from other members of my faith.
15. **T/F** I provide support to other members of my faith.

Introspection

A. If you were trapped alone on a deserted island for a period of time, could you still worship as you would like?

B. Why?

C. Why not?

D. What is it about the building, other members, and associations that makes or does not make religion significant to you?

Key

This assessment is biased toward religiosity, the higher your score the higher your religiosity.

1–15 = T. 1–5 measure formal religious participation. 6–10 measure private religious participation. 11–15 measure religious social connectedness. After having taken this assessment would you still place yourself in the same place on the continuum? How do you compare to your partner?

Your Partner's Religiosity Continuum

This assessment is designed to help your partner discover his or her levels of religiosity or participation in formal religion and its meaning in his or her personal life. **Partner:** First, write your initials on the continuum below in the place that best represents your overall religiosity. Then answer the questions T (True) or F (False) below as accurately as possible.

Very Religious ——————————————— Neutral ——————————————— Not Very Religious

Questions

1. T/F I consider myself to be a member of a religious denomination.
2. T/F I attend worship services regularly throughout the year.
3. T/F I consider myself to be an "active" member.
4. T/F I regularly volunteer time to religious activities.
5. T/F I regularly donate money to my religion.
6. T/F I worship even when I'm not in formal worship services (i.e.: at home, work, etc. . . .).
7. T/F I often meditate upon my beliefs.
8. T/F I read sacred writings on my own.
9. T/F I pray regularly.
10. T/F I live my beliefs.
11. T/F I believe in a "higher power."
12. T/F I need the influence of a "higher power."
13. T/F I enjoy the relationships I have with other members of my faith.
14. T/F I receive support from other members of my faith.
15. T/F I provide support to other members of my faith.

Introspection

A. If you were trapped alone on a deserted island for a period of time, could you still worship as you would like?

B. Why?

C. Why not?

D. What is it about the building, other members, and associations that makes or does not make religion significant to you?

Key

This assessment is biased toward religiosity, a higher score implies higher religiosity.

1–15=T. 1–5 measure formal religious participation. 6–10 measure private participation. 11–15 measure religious social connectedness. After having taken this assessment would you still place yourself in the same place on the continuum? How do you compare to your partner?

Sexual Harassment Awareness Quiz

Please answer T (True) or F (False) for the questions below as they apply to your perception of today's workplace environment.

1. **T/F** A casual touch of another person's body is to be expected in the workplace.
2. **T/F** Sexual rumors about a coworker are acceptable in the workplace.
3. **T/F** We should come to expect being brushed up against in a sexual way.
4. **T/F** Everybody makes sexual remarks or suggestions. It's O.K.
5. **T/F** Conversations that are too personal are acceptable.
6. **T/F** When coworkers hang up pornographic pictures in the workplace it is their right.
7. **T/F** We are paid to put up with unwelcome whistles, gestures, or noises.
8. **T/F** It is O.K. for others to tell dirty jokes at work, even if it bothers us.
9. **T/F** When a person does an obscene gesture, it's best to ignore it.
10. **T/F** Mooning and flashing are done in jest. They are acceptable.
11. **T/F** Displaying sexual quotes, objects, or stories is a form of sexual harassment.
12. **T/F** It is the employers' and employees' responsibility to create a harassment-free workplace environment.
13. **T/F** When a superior offers promotions, favors, opportunities in return for sexual favors, that is sexual harassment.
14. **T/F** When superiors or coworkers behave in offensive ways at work and do not respond to your request to be more professional, that is another form of sexual harassment called a hostile work environment.
15. **T/F** Usually, it takes more than one offense before a workplace is considered "hostile."
16. **T/F** Women sometimes harass other women and men other men.
17. **T/F** Women can be harassers of men in the workplace.
18. **T/F** Treating a sexual harassment complaint seriously is the best way to respond.
19. **T/F** Victims of harassment should immediately go to superiors or coworkers to report it.
20. **T/F** Keeping track of events through personal notes is a good idea with harassment.

Key

1–10 = false, 11–20 = true.

Give yourself **1 point** for each correct answer. The higher your score the better you understand the finer aspects of working in the modern workplace.

In today's workplace men and women must be skilled in knowing their rights and responsibilities pertaining to a sexual harassment-free workplace. We first must understand what it is, why it is illegal, and why it can be destructive in the workplace. Next we must learn what our role as workers and managers is in creating a safe and healthy environment at work. Does your school, place of employment, or business have a sexual harassment policy/program? Check it out!

Do the findings from this assessment accurately reflect you, your expectations, and your life experience? Why?

Quality of Health Care Coverage

In this assessment, answer questions based on the health care coverage you had in your family of origin or in your current family (pick only the most relevant one). Answer each question T (True) or F (False) as accurately as possible. Do not look at the key until after the assessment is completed.

1. T/F   I have had adequate health care coverage most of my life.
2. T/F   I am rarely sick.
3. T/F   I am comfortable talking to my doctor.
4. T/F   I sometime ask my doctor questions if I disagree with him/her.
5. T/F   Financially, we can take our children to the doctor at any time.
6. T/F   Our copays are too high.
7. T/F   Our insurance does not cover the major medical problems adequately.
8. T/F   I have had to use welfare medical coverage in my life.
9. T/F   It could break us financially if one of us gets seriously ill.
10. T/F   I haven't seen the dentist in years because it costs too much.
11. T/F   If people want health coverage, they'd better be willing to pay for it.
12. T/F   I've gone for weeks without seeking medical help for something that was bothering me.
13. T/F   I sometimes don't have enough money to pay for my medicines.
14. T/F   At least one person in the family is on a life-long medication program.
15. T/F   If I had an accident requiring some medical help, I don't know who I'd go to.
16. T/F   It's alright for other family members to get sick but I can't.
17. T/F   Medical coverage ought to be available to every one in this country.
18. T/F   If I had been covered better, then I would be healthier today.
19. T/F   We put off going to the doctor as long as possible.
20. T/F   I have few rights as a patient.
21. T/F   I get a physical exam every year.
22. T/F   My insurance company won't pay for care that I deserve.
23. T/F   Doctors simply get overpaid for what they do.
24. T/F   I often use natural medical interventions.
25. T/F   I use massage therapy.
26. T/F   I use chiropractic therapy.
27. T/F   I use pharmaceutical interventions (medications).
28. T/F   Doctors don't have all of the answers.
29. T/F   No one has all of the medical answers.
30. T/F   Medical expenses eat up our budget.
31. T/F   Overall our medical coverage exceeds our needs.
32. T/F   Sometimes I use a healer rather than a trained medical type.
33. T/F   My doctor and nurse sometimes treat me disrespectfully.
34. T/F   Lawsuits against medical providers bring balance to the medical system.
35. T/F   Overall I'm a well person.

Key

1 = T, 2 = T, 3 = T, 4 = T, 5 = T, 6 = F, 7 = F, 8 = F, 9 = F, 10 = F, 11 = F, 12 = F, 13 = F, 14 = F, 15 = F, 16 = F, 17 = T, 18 = F, 19 = F, 20 = F, 21 = T, 22 = F, 23 = T, 24 = T, 25 = T, 26 = T, 27 = T, 28 = T, 29 = T, 30 = F, 31 = T, 32 = T, 33 = F, 34 = F, 35 = T

Your Score _____

This assessment is biased toward adequate, healthy, and balanced health care coverage. A score of 35 indicates very balanced health care coverage. A score of 0 indicates very unhealthy coverage. Most people would score somewhere in between. Do you have any concerns about issues associated with the questions above? Discuss them with a family member, close friend, or health care provider. Are there choices you can make to improve upon the situation?

The Value of Education in My Family

Self-Version

A nswer the questions below T (True) or F (False) using your family of origin as the reference. Be as accurate as possible and look at the key only after completing the assessment. Have your partner, close friend, or family member take the Partner Version.

1. **T/F** My grandparent(s) attended college.
2. **T/F** My parent(s) attended college.
3. **T/F** I attended college.
4. **T/F** I hope my children attend college.
5. **T/F** We have professionally trained people in our family.
6. **T/F** My grandparents are middle class or above.
7. **T/F** My parents are middle class or above.
8. **T/F** We are middle class or above.
9. **T/F** I hope my children are middle class or above.
10. **T/F** Our family tends to help out financially if someone wants to go to college.
11. **T/F** In our home learning was deeply valued.
12. **T/F** We were sent to pre-kindergarten classes.
13. **T/F** I received help with school work while growing up.
14. **T/F** We received private educational training while growing up.
15. **T/F** There are more high school graduates than drop outs among my brothers and sisters.
16. **T/F** Education means better jobs.
17. **T/F** Education helps you see things more clearly.
18. **T/F** Education is the door to success.
19. **T/F** My relatives come to my graduations.
20. **T/F** I received presents for graduation.
21. **T/F** My relatives want me to become more educated.
22. **T/F** A high school diploma impresses my relatives.
23. **T/F** A college degree impresses my relatives.
24. **T/F** A graduate degree impresses my relatives.
25. **T/F** I get lots of support in college and do not have to work while being a student.

Key

1–25 = True

Your Score _____

A higher score indicates a higher family value on education.

Fill out the information requested on the following family members:

Mother's Information	Father's Information
Mother's education _____ and income $ _____	Father's education _____ and income $ _____
Grandmother's education _____ and income $ _____	Grandmother's education _____ and income $ _____
Grandfather's education _____ and income $ _____	Grandfather's education _____ and income $ _____
Other relatives' education levels and incomes	Other relatives' education levels and incomes

By examining the relatives on your mother's and father's side, you can see if a relationship between education and income emerges. Typically, the higher the education then the higher the income. Is this true in your family? How far do you plan to go in your education? Have your partner do the partner assessment and then discuss the results.

The Value of Education in My Family

Partner Version

Answer the questions T (True) or F (False) below using your family of origin as the reference. Be as accurate as possible and look at the key only after completing the assessment. Have your partner, close friend, or family member take the Self-Version.

---

1. **T/F**   My grandparent(s) attended college.
2. **T/F**   My parent(s) attended college.
3. **T/F**   I attended college.
4. **T/F**   I hope my children attend college.
5. **T/F**   We have professionally trained people in our family.
6. **T/F**   My grandparents are middle class or above.
7. **T/F**   My parents are middle class or above.
8. **T/F**   We are middle class or above.
9. **T/F**   I hope my children are middle class or above.
10. **T/F**   Our family tends to help out financially if someone wants to go to college.
11. **T/F**   In our home learning was deeply valued.
12. **T/F**   We were sent to pre-kindergarten classes.
13. **T/F**   I received help with school work while growing up.
14. **T/F**   We received private educational training while growing up.
15. **T/F**   There are more high school graduates than drop outs among my brothers and sisters.
16. **T/F**   Education means better jobs.
17. **T/F**   Education helps you see things more clearly.
18. **T/F**   Education is the door to success.
19. **T/F**   My relatives come to my graduations.
20. **T/F**   I received presents for graduation.
21. **T/F**   My relatives want me to become more educated.
22. **T/F**   A high school diploma impresses my relatives.
23. **T/F**   A college degree impresses my relatives.
24. **T/F**   A graduate degree impresses my relatives.
25. **T/F**   I get lots of support in college and do not have to work while being a student.

Key

1–25 = True

Your Score _____

A higher score indicates a higher family value on education.

Fill out the information requested on the following family members:

Mother's Information	Father's Information
Mother's education _____ and income $ _____	Father's education _____ and income $ _____
Grandmother's education _____ and income $ _____	Grandmother's education _____ and income $ _____
Grandfather's education _____ and income $ _____	Grandfather's education _____ and income $ _____
Other relatives' education levels and incomes	Other relatives' education levels and incomes

By examining the relatives on your mother's and father's side, you can see if a relationship between education and income emerges. Typically, the higher the education then the higher the income. Is this true in your family? How far do you plan to go in your education? Have your partner do the self-assessment and then discuss the results.

Net Worth Data Sheet

In the next two financial worksheets, fill in the data according to your current financial circumstances. Then fill in your 1-, 5-, and 10-year goals for each category.

Assets	Now (Expressed in Current Value)	1-Year Goal	5-Year Goal	10-Year Goal
Cash				
Bank				
Checking				
Savings				
Certificates				
Savings Bonds				
Other Bonds				
Life Insurance				
Mutual Funds				
Retirement Plans				
Private Pension Plan				
Profit Sharing Plan				
Home Mortgage—Market Value				
Other Real Estate				
Car(s) Loans—Market Value				
Furniture				
Large Appliances				
Antiques/Art				
Jewelry				
Silverware				
Stamp/Coin Collection				
Debts Others Owe You				
Other				
Total Assets				

Liabilities	Now (Expressed in Current Value)	1-Year Goal	5-Year Goal	10-Year Goal
Current Bills				
Charge Accounts				
Credit Cards				
Taxes				
Installments				
Mortgages				
Other				
Total Liabilities				
Net Worth (total assets minus total liabilities)				

Your Intergenerational Mobility

Fill out the socioeconomic status of yourself, your parents, and your grandparents (estimate where needed). List the occupation, education level, and economic class (upper, middle, or lower) for each person. What has kept things the same, and what has caused them to improve or decline between you and your previous generations?

Third Generation	Second Generation	First Generation
		Your Father's Father's Name: Occupation: Education Level: Class:
Your Name: Occupation: Education Level: Class:	Your Father's Name: Occupation: Education Level: Class:	Your Father's Mother's Name: Occupation: Education Level: Class:
	Your Mother's Name: Occupation: Education Level: Class:	Your Mother's Father's Name: Occupation: Education level: Class:
		Your Mother's Mother's Name: Occupation: Education Level: Class:

Your Spouse's Intergenerational Mobility

Fill out the socioeconomic status of your spouse, spouse's parents, and grandparents (estimate where needed). List the occupation, education level, and economic class (upper, middle, or lower) for each person. What has kept things the same, and what has caused them to improve or decline between your spouse and previous generations? What will this marriage do to change things?

Third Generation	Second Generation	First Generation
		Spouse's Father's Father's Name: Occupation: Education Level: Class:
Spouse's Name: Occupation: Education Level: Class:	Spouse's Father's Name: Occupation: Education Level: Class:	Spouse's Father's Mother's Name: Occupation: Education Level: Class:
	Spouse's Mother's Name: Occupation: Education Level: Class:	Spouse's Mother's Father's Name: Occupation: Education level: Class:
		Spouse's Mother's Mother's Name: Occupation: Education Level: Class:

Your Intragenerational Mobility

Fill out the socioeconomic status of yourself and all of your siblings and step siblings (estimate where needed). List the occupation, education level, and economic class (upper, middle, or lower), and current annual income in dollars for each person. In this exercise you are comparing your status to that of your brothers and sisters.

Name in Birth Order	Current Occupation	Current Education Level	Economic Class	Current Estimated Income $
1.				
2.				
3.				
4.				
5.				
6.				
7.				
8.				
9.				

1. Why are there similarities?

2. Why are there differences?

3. Has your or your sibling's marital status (i.e.: married, never married, or divorced) played a part in the answers to questions 1 or 2?

4. What opportunities were available to some children but not to others?

Your Spouse's Intragenerational Mobility

Fill out the socioeconomic status of your spouse and all of his/her siblings and step siblings (estimate where needed). List the occupation, education level, and economic class (upper, middle, or lower), and current annual income in dollars for each person. In this exercise you are comparing your spouse's status to that of his/her brothers and sisters.

Name in Birth Order	Current Occupation	Current Education Level	Economic Class	Current Estimated Income $
1.				
2.				
3.				
4.				
5.				
6.				
7.				
8.				
9.				

1. Why are there similarities?

2. Why are there differences?

3. Has your spouse's or his/her sibling's marital status (i.e.: married, never married, or divorced) played a part in the answers to questions 1 or 2?

4. What opportunities were available to some children but not to others?

My Social and Financial Position

Where do I fall on the social/financial scale?
Am I:

Wealthy?

Upper class?

Upper middle class?

Lower middle class?

Poor?

In poverty?

How has my family and community affected my position in society?

Many beginning families start out in poverty. They are poor about 12% of their life and then move out of poverty. Certainly young families in college may be considered poor, but after training and placement in employment, they move up in their financial status.

How do you see your status related to being poor?

Does poverty automatically dictate unhappiness? Can a person be poor and still be happy?

What are my goals for financial stability?

1.

2.

3.

4.

What are my partner's goals for financial stability?

1.

2.

3.

4.

Our goals together:

1.

2.

3.

Who Are the Poor in America?

Children are the poor in America. Single mothers with less than a high school degree are the main persons in poverty. Early sexuality or sexualization of a young girl defeats her potential for later education and future equality.

List your ideas of how to combat this dilemma:

1.

2.

3.

4.

5.

How can women gain equality and autonomy in society?

1.

2.

How can women gain equality and autonomy in the family?

1.

2.

What should be expected of fathers of children to single mothers? What would you do?

How do your ideas affect children who are in poverty for positive change?

1.

2.

Social Reproduction

ocial reproduction is when children grow up to continue in the same class position as their parents. Children adopt the values, religion, culture, dress, dialogue, and recreational preferences of their family and friends.

Have I followed in the same class position as my parents? If yes, list the areas of similarities. If no, list those ways.

Yes:

1.

2.

3.

4.

5.

No:

1.

2.

3.

4.

5.

How do I want my own family to be? Do I want to follow in my parents' footsteps or change?

Partner's response:

What do I want to change?

How do I change?

The Self-Portrait:
An Exercise in Appreciating
Our Self and Our Groups

*(**Note:** Do Not Read Understanding the Self- and Super Self-Portrait Until After Having Completed the Self- and Super Self-Portraits.)*

You will need a box of crayons (yes crayons) for this project. As long as there are enough colors to give variety, just about any brand of crayon will suffice for this project. You will be an artist drawing a picture of your face, which represents how you see yourself, not necessarily how you actually look. Your self-portrait will reflect your heritage and tradition as well as your biological ancestry.

Step 1 Draw a picture of yourself
Your drawing should be done on the blank page provided. Since none of us are truly white, black, yellow, or red, and so on, you can experiment with the crayons and get the colors that best fit how you see yourself. Draw yourself as you perceive you to be. This is not an exercise in art. It can be a very simple drawing but you should apply careful thought and give attention to the colors you choose.

Step 2 Describe yourself
When the drawing is finished, write 20 descriptive words about yourself along the margins of the paper: 5 to describe your personality and likes/dislikes; 5 to describe your ancestral heritage; 5 about what you hope to be in the future; and 5 more about your prejudices and difficulties in interacting with other people.

Step 3 Explain your work to another person
With your drawing complete, and your descriptive words in place, you should now explain what you did and what you have written and why. Allow someone you know to ask you questions and respond to them as honestly as possible. If you did this activity with someone else, then let them explain their work to you.

The Super Self-Portrait:
An Exercise in Appreciating Other Groups

*(**Note:** Do Not Read Understanding the Self- and Super Self-Portrait Until After Having Completed the Self- and Super Self-Portraits)*

For this activity, you will again need a box of crayons. You will again be an artist drawing a picture. But this time from a significantly different point of view. Think of how you might look if you could be any race, ethnicity, color, creed, or national origin you wanted to be **but not the one you currently are.** *(Do not reduce this exercise to a choice between your present race and some other.)* It is *not* a question of "would you change your race or anything else." It *is* basically an exercise asking you to depict yourself as a member of some race or ethnicity that you have admired during your life.

Step 1 Chose a group
At this point, you should chose to draw yourself as a member of a group other than who you are. Choose a group that you have always admired or romanticized. Imagine how you might appear if you had been born to parents who belong to that group.

Step 2 Draw a picture of yourself
Use the blank page provided. Again, get the colors that best fit how you want yourself to appear. This may be a very simple drawing. It will be enhanced when you give careful thought and attention to the colors and descriptive words.

Step 3 Describe yourself
Write 20 descriptive words about your hypothetical super self along the margins of the paper: 5 about what your personality and likes/dislikes would be had you been born in the group you chose; 5 about your new ancestral heritage; 5 about what you might be in the future; and 5 more about your *strengths and skills* in interacting with other people. Again, this is not an exercise in art.

Step 4 Explain your work to another person
With your drawing complete, and your descriptive words in place, you should now explain what you did and what you have written and why. Allow others to ask you questions and respond to them as honestly as possible. If you did this activity with someone else, then let them explain their work to you.

Understanding the Self- and Super Self-Portrait

*(**Note:** Do not read this page until after you have completed the Self- and Super Self-Portraits. By reading this page first, you may bias your expression and interfere with the purpose of the project.)*

The Self-Portrait is designed to allow for self-observation, description, and criticism. It allows you to safely and pleasantly approach how you appear, who you are, where you come from, what you would dream to do, and some of your weaknesses in relationships. By sharing it you get to actually verbalize these observations to another person. Many find that the explanation and interactions are the best part of the activity.

The Super Self-Portrait is designed to develop our appreciation of other groups. It allows us to be very personal as we explore in depth the individual we might be if we were an "other." This project draws heavily on our categorical thinking. But it is valuable because it allows us to circumvent a very deeply embedded cultural roadblock to self-discovery. You see, in most cultures, we aren't really allowed to openly discuss our own good qualities. It sounds too much like bragging. But by suggesting to you that you are drawing a hypothetical you, born in another culture, you are allowed to project your positive qualities safely and without shame. Typically the descriptive words chosen for the Super Self-Portrait almost always best describe the person who wrote them.

By doing the Self-Portrait, we discover more clearly how we perceive ourselves to be (strengths and weaknesses). By doing the Super Self-Portrait, we overcome cultural inhibitions to clearly viewing our strengths. At the same time we gain experience in appreciating people from different cultures. Studies on relationships between racial and ethnic group members indicate that we are more attracted to and therefore more likely to get along with those who appear to be similar to us. These assessments allow you to see deep similarities in other groups and they establish appreciation patterns at the same time. You also discover that you are similar in many ways to people whom you sometimes view as being different. What did you learn about yourself and others in this exercise?

My Predispositions to Prejudice

Self-Version

In this assessment simply answer T (True) or F (False) if any of these questions apply. All people are more or less disposed to prejudiced thinking—some more than others. This assessment will help you to identify any patterns of thinking, known to be associated with prejudice. It does not tell you if your are in fact a prejudiced person. It merely identifies patterns associated with being prejudiced. Do not look at the key until after completing the assessment. Have your partner take the Partner Version and discuss the results.

---

1. T/F   I tend to like most people.
2. T/F   I have many friends.
3. T/F   I get along well with most people.
4. T/F   I enjoy the uniqueness of different groups of people.
5. T/F   I was hurt deeply by someone in the past. Now all people like that bother me.
6. T/F   When I grew up, people in my family usually had something bad to say about certain groups of people.
7. T/F   Sometimes I feel cheated because certain groups of people are taking away my livelihood.
8. T/F   I was afraid of certain groups of people when I was a child.
9. T/F   People will get along if they need each other.
10. T/F   I don't like myself that much.
11. T/F   I have been the victim of discrimination in my life.
12. T/F   I have been the victim of prejudice in my life.
13. T/F   I have no bad feelings toward any group of people today.
14. T/F   There are certain types of people I simply can't stand to be around.
15. T/F   My parents never let me tell my point of view.
16. T/F   My parents were far too controlling.
17. T/F   My mother and/or father absolutely hate certain groups of people.
18. T/F   If I got angry at my parents, then they'd listen and we worked it out.
19. T/F   My parents have prejudices.
20. T/F   My grandparents have prejudices.
21. T/F   If certain people were not living here, then there would be more resources for those of us who should live here.
22. T/F   All people can get along.
23. T/F   I used to be picked on a lot.
24. T/F   I can't stand people like myself.
25. T/F   I have been violently attacked in my life.
26. T/F   My friends tend to be prejudiced.
27. T/F   Some groups of people are simply more capable than others.

28. T/F Certain groups of people bring out strong negative emotions in me.
29. T/F Certain groups of people tend to be rich and snobby.
30. T/F Sometimes, when I meet someone new, I have a strong negative reaction to them for no reason.
31. T/F I struggle with my own prejudices.
32. T/F Sometimes pressure builds up in me and I have to take it out on someone.
33. T/F I used to be a bully.
34. T/F Sometimes certain groups of people just do stupid things.
35. T/F I fit in.
36. T/F Everyone has some prejudices.

Key

1 = T, 2 = T, 3 = T, 4 = T, 5 = F, 6 = F, 7 = F, 8 = F, 9 = T, 10 = F, 11 = F, 12 = F, 13 = T, 14 = F, 15 = F, 16 = F, 17 = F, 18 = T, 19 = T, 20 = T, 21 = F, 22 = T, 23 = F, 24 = F, 25 = F, 26 = F, 27 = F, 28 = F, 29 = F, 30 = F, 31 = T, 32 = F, 33 = F, 34 = F, 35 = T, 36 = T

Your Score _____

This assessment is biased toward open-minded, teachable, and nonprejudicial thinking patterns. Getting 36 incorrect answers does *not* indicate prejudice any more than 1 incorrect answer would. This assessment only provides insights into patterns that may be prejudiced. Let your partner, family member, or friend take the partner assessment and talk about the results. If you suspect prejudice at any level keep in mind the principle of change—people can change. Most people have some level of prejudice. Many are successful at acknowledging it and managing it wisely. What do you think?

My Predispositions to Prejudice

Partner Version

In this assessment simply answer T (True) or F (False) if any of these questions apply. All people are more or less disposed to prejudiced thinking—some more than others. This assessment will help you to identify any patterns of thinking, known to be associated with prejudice. It does not tell you if your are in fact a prejudiced person. It merely identifies patterns associated with being prejudiced. Do not look at the key until after completing the assessment. Have your partner take the Self-Version and discuss results.

1. T/F I tend to like most people.
2. T/F I have many friends.
3. T/F I get along well with most people.
4. T/F I enjoy the uniqueness of different groups of people.
5. T/F I was hurt deeply by someone in the past. Now all people like that bother me.
6. T/F When I grew up, people in my family usually had something bad to say about certain groups of people.
7. T/F Sometimes I feel cheated because certain groups of people are taking away my livelihood.
8. T/F I was afraid of certain groups of people when I was a child.
9. T/F People will get along if they need each other.
10. T/F I don't like myself that much.
11. T/F I have been the victim of discrimination in my life.
12. T/F I have been the victim of prejudice in my life.
13. T/F I have no bad feelings toward any group of people today.
14. T/F There are certain types of people I simply can't stand to be around.
15. T/F My parents never let me tell my point of view.
16. T/F My parents were far too controlling.
17. T/F My mother and/or father absolutely hate certain groups of people.
18. T/F If I got angry at my parents, then they'd listen and we worked it out.
19. T/F My parents have prejudices.
20. T/F My grandparents have prejudices.
21. T/F If certain people were not living here, then there would be more resources for those of us who should live here.
22. T/F All people can get along.
23. T/F I used to be picked on a lot.
24. T/F I can't stand people like myself.
25. T/F I have been violently attacked in my life.
26. T/F My friends tend to be prejudiced.

27. T/F   Some groups of people are simply more capable than others.
28. T/F   Certain groups of people bring out strong negative emotions in me.
29. T/F   Certain groups of people tend to be rich and snobby.
30. T/F   Sometimes, when I meet someone new, I have a strong negative reaction to them for no reason.
31. T/F   I struggle with my own prejudices.
32. T/F   Sometimes pressure builds up in me and I have to take it out on someone.
33. T/F   I used to be a bully.
34. T/F   Sometimes certain groups of people just do stupid things.
35. T/F   I fit in.
36. T/F   Everyone has some prejudices.

Key

1 = T, 2 = T, 3 = T, 4 = T, 5 = F, 6 = F, 7 = F, 8 = F, 9 = T, 10 = F, 11 = F, 12 = F, 13 = T, 14 = F, 15 = F, 16 = F, 17 = F, 18 = T, 19 = T, 20 = T, 21 = F, 22 = T, 23 = F, 24 = F, 25 = F, 26 = F, 27 = F, 28 = F, 29 = F, 30 = F, 31 = T, 32 = F, 33 = F, 34 = F, 35 = T, 36 = T

Your Score _____

This assessment is biased toward open-minded, teachable, and nonprejudicial thinking patterns. Getting 36 incorrect answers does *not* indicate prejudice any more than 1 incorrect answer would. This assessment only provides insights into patterns that may be prejudiced. Let your partner, family member, or friend take the partner assessment and talk about the results. If you suspect prejudice at any level keep in mind the principle of change—people can change. Most people have some level of prejudice. Many are successful at acknowledging it and managing it wisely. What do you think?

Understanding Problems at the Group Level:
An Exercise in Overcoming Prejudice

I n this exercise, you will have an opportunity to better understand prejudice between you and certain groups of people. This is a very personal type assessment and it is recommended that you *not* talk to anyone about it once it's completed. It is designed to help you personally, as an individual. The questions are very personal and go to the core of our sensitive values, beliefs, emotions, and predispositions. Begin by thinking for a day or so about a group of people: racial, religious, ethnic, lifestyle, physical features, skin color, and so on, that have been particularly difficult for you to accept or tolerate. Pick only one group and pick the one that has the most intensity of effect on you. Walk through these questions thoughtfully and reflect upon your answers as you go.

1. List the group here:_____

2. What is your earliest recollection of interacting with, knowing about, or being concerned with this group? Brief paragraph.

3. Did you have any direct interactions with members of this group? If yes, list the significant ones in either positive or negative context.

   Positive:

   Negative:

4. Did family, friends, teachers, or other authority figures shape your attitude about these group members? If so how?

5. Who else do you know with similar attitudes about this group? List names.

6. What **emotions** arise when you think about, interact with, or hear about this group? List them in the context they occurred.

7. What **thoughts** come to you when you think about, interact with, or hear about this group? List them in the context they occurred.

8. List how the following people would describe members of this group or the group as a whole:

Mother:

Father:

Grandmother:

Grandfather:

Closest friend:

Expert in the field:

9. Pretend for the sake of understanding that you hire people for a major manufacturing company. You have three applicants for a very good paying high status job. All three have the same experience, education level, interpersonal skills, and in every way are equally qualified. The only difference is that one of them is a member of the group you are focusing on in this exercise. After all is said and done, no one will know why you hired who you hired and legally your paperwork will not indicate discrimination.

Who gets the job, the group member or one of the other two candidates?

List any feelings associated with your choice:

List any thoughts associated with your choice:

Was your decision fair to the other two candidates as well as to the group member?

Sometimes when we have issues with a group, we are predisposed to act either favorably or unfavorably toward its members. When we meet them for the first time those predispositions show up.

10. Has any one ever acted favorably toward or unfavorably against you because of your group membership? Explain.

Does that have an effect on decisions like the one you made in question 9?

11. Place an X on the continuum below indicating where you might be in terms of being prejudiced about this group.

Not at all prejudiced ————————————————————————————————————— Very prejudiced

Final Thoughts

Step 1   *Acknowledging a problem in ourselves is the very first step in changing ourselves. If you are prejudiced it is necessary to acknowledge it as well. These other steps help in the process of managing and overcoming your prejudices:*

Step 2   *Look for its origins in your past;*

Step 3   *Be teachable and willing to change;*

Step 4   *Set specific goals to manage it;*

Step 5   *Gently confront prejudice when we see it;*

Step 6   *Make it a life-long pursuit.*

Your Family Race, Ethnic, and Cultural Heritage

In this worksheet, focus on the components of your heritage that make you and your family unique and that provide rich diversity to your life experience. Consult other family members when needed. By reporting the racial, ethnic, and/or cultural heritages for your nuclear and extended family you may better discover how your current family came to be what it is today and how those heritages may enrich your life. You may also do this for your partner.

Heritage Factors	My Family of Origin	Father's Family of Origin	Mother's Family of Origin	Spouse's Family of Origin	Father-in-Law's Family of Origin	Mother-in-Law's Family of Origin
National Origin						
Religion						
Language Spoken						
Economic Class						
Work Experience						
List Three Main Family Traditions	1. 2. 3.	1. 2. 3.	1. 2. 3.	1. 2. 3.	1. 2. 3.	1. 2. 3.
How Important Was Family Heritage?						
Role of Women in Family						
Dinner Table Activities						
Holidays Celebrated						
Other Factors						

Now that you have identified these heritage factors, discuss them with your spouse, partner, or family. What type of heritage will be emphasized in your family of marriage/cohabitation/procreation? Are there dormant traditions or values held by previous generations that you would like to revive? Why or why not?

Your Mother's, Father's, and Your Own Ancestral/Biological Heritage

In the following pages, fill out your mother's and then your father's Ancestral/Biological Heritage sheets. These will provide you with details about your heritage that extend back to your great grandparents on both your mother's and father's side. Once you've completed filling out both sheets as completely as possible, answer the questions below. They will help to clarify how your heritage, as it was passed down to you, has influenced your own definition of your diverse self.

Questions

1. How strongly did your father identify with his racial and or ethnic group?

2. What was the role of his religion, language, and marriage in that self-identification?

3. What influence did his parents and/or grandparents have in that self-identification?

4. How strongly did your mother identify with his racial and/or ethnic group?

5. What was the role of her religion, language, and marriage in that self-identification?

6. What influence did her parents and/or grandparents have in that self-identification?

7. How strongly do you identify with your racial and/or ethnic group?

8. What is the role of religion, language, and/or marriage in that self-identification?

9. What influence did/do your parents, grandparents, and/or great grandparents have in that self-identification?

10. Is there a remote racial, religious, or ethnic heritage that you more strongly identify with than other heritages? Why or why not? If so what is it?

It may become a meaningful experience to talk to parents and other relatives about this assignment. Compare your views and feelings to theirs and share why these views are important to you. Think about what you learned from them that you did not know before. Will your children know such things about you?

My Mother's Ancestral/Biological Heritage

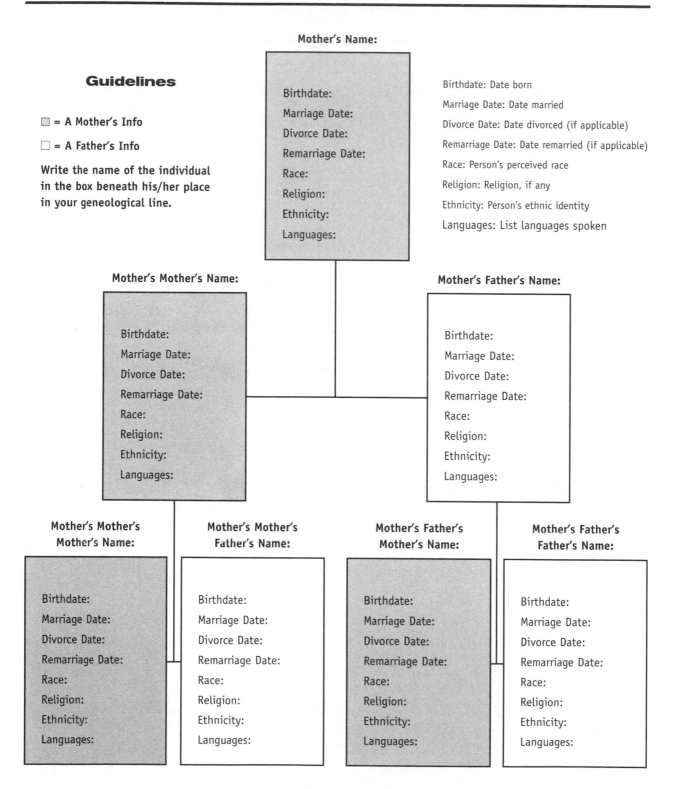

Guidelines

▨ = A Mother's Info

▢ = A Father's Info

Write the name of the individual in the box beneath his/her place in your geneological line.

Mother's Name:

Birthdate:

Marriage Date:

Divorce Date:

Remarriage Date:

Race:

Religion:

Ethnicity:

Languages:

Birthdate: Date born

Marriage Date: Date married

Divorce Date: Date divorced (if applicable)

Remarriage Date: Date remarried (if applicable)

Race: Person's perceived race

Religion: Religion, if any

Ethnicity: Person's ethnic identity

Languages: List languages spoken

Mother's Mother's Name:

Birthdate:

Marriage Date:

Divorce Date:

Remarriage Date:

Race:

Religion:

Ethnicity:

Languages:

Mother's Father's Name:

Birthdate:

Marriage Date:

Divorce Date:

Remarriage Date:

Race:

Religion:

Ethnicity:

Languages:

Mother's Mother's Mother's Name:

Birthdate:

Marriage Date:

Divorce Date:

Remarriage Date:

Race:

Religion:

Ethnicity:

Languages:

Mother's Mother's Father's Name:

Birthdate:

Marriage Date:

Divorce Date:

Remarriage Date:

Race:

Religion:

Ethnicity:

Languages:

Mother's Father's Mother's Name:

Birthdate:

Marriage Date:

Divorce Date:

Remarriage Date:

Race:

Religion:

Ethnicity:

Languages:

Mother's Father's Father's Name:

Birthdate:

Marriage Date:

Divorce Date:

Remarriage Date:

Race:

Religion:

Ethnicity:

Languages:

My Father's Ancestral/Biological Heritage

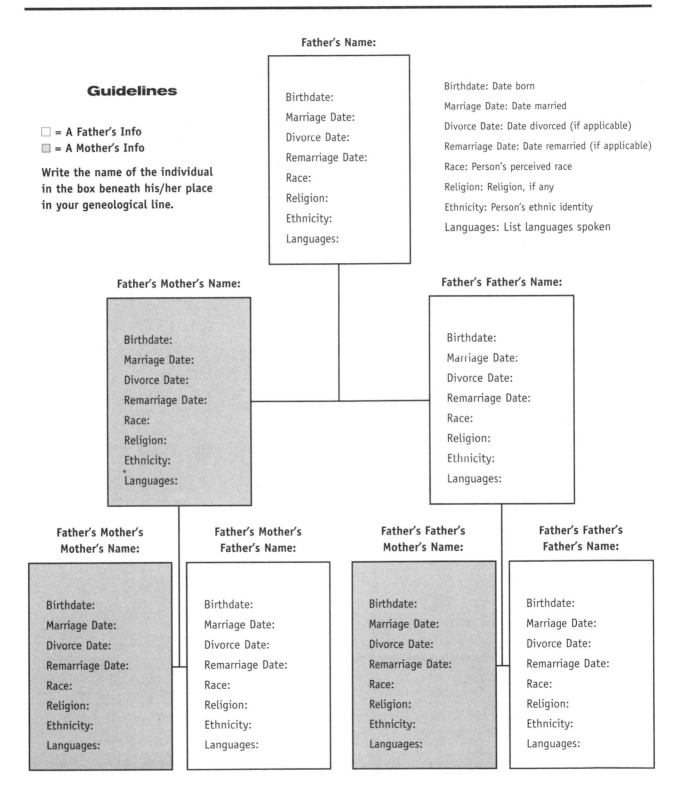

Father's Name:

Birthdate:

Marriage Date:

Divorce Date:

Remarriage Date:

Race:

Religion:

Ethnicity:

Languages:

Guidelines

☐ = A Father's Info
▨ = A Mother's Info

Write the name of the individual in the box beneath his/her place in your geneological line.

Birthdate: Date born

Marriage Date: Date married

Divorce Date: Date divorced (if applicable)

Remarriage Date: Date remarried (if applicable)

Race: Person's perceived race

Religion: Religion, if any

Ethnicity: Person's ethnic identity

Languages: List languages spoken

Father's Mother's Name:

Birthdate:

Marriage Date:

Divorce Date:

Remarriage Date:

Race:

Religion:

Ethnicity:

Languages:

Father's Father's Name:

Birthdate:

Marriage Date:

Divorce Date:

Remarriage Date:

Race:

Religion:

Ethnicity:

Languages:

Father's Mother's Mother's Name:

Birthdate:

Marriage Date:

Divorce Date:

Remarriage Date:

Race:

Religion:

Ethnicity:

Languages:

Father's Mother's Father's Name:

Birthdate:

Marriage Date:

Divorce Date:

Remarriage Date:

Race:

Religion:

Ethnicity:

Languages:

Father's Father's Mother's Name:

Birthdate:

Marriage Date:

Divorce Date:

Remarriage Date:

Race:

Religion:

Ethnicity:

Languages:

Father's Father's Father's Name:

Birthdate:

Marriage Date:

Divorce Date:

Remarriage Date:

Race:

Religion:

Ethnicity:

Languages:

How Did You Meet Each Other?

Answer the questions below in reference to your current relationship and all of the past significant relationships you can recall. This assessment will clarify for you how you tend to meet someone and the patterns associated with the beginnings of your relationships.

List the first names of all the significant relationships you want to consider for this assessment:

1.	4.	7.	10.
2.	5.	8.	11.
3.	6.	9.	12.

Where did you meet each person?

1.	4.	7.	10.
2.	5.	8.	11.
3.	6.	9.	12.

Who introduced you?

1.	4.	7.	10.
2.	5.	8.	11.
3.	6.	9.	12.

What special circumstances existed if any?

1.	4.	7.	10.
2.	5.	8.	11.
3.	6.	9.	12.

What main factors did you share in common with each person?

1.	4.	7.	10.
2.	5.	8.	11.
3.	6.	9.	12.

Look at your answers for each question. Do patterns emerge such as:

1. Where do you tend to meet significant others?
2. Is someone else typically involved in terms of introducing you?
3. What type of circumstances are typically involved?
4. In terms of having things in common, what role did homogamy play in being in the same place, being introduced by third party (also similar to you), and having similar interests that drew you to the same place?
5. How did homogamy, or the lack thereof, contribute to the outcome of your relationship?

Intimacy Assessment Questionnaire

This assessment is designed to help you become more aware of intimacy and how it occurs in your life. Answer the questions T (True) or F (False) below by circling the answer that best fits your relationships, experience, and views. Wait until you have completed the questionnaire before consulting the key.

1. T/F I rarely share my true self with others.
2. T/F Sometimes we get mad at one another and it is a sure sign that we have fallen out of love.
3. T/F I was basically happier years ago than I am now.
4. T/F When I am in a very close relationship I feel comfortable.
5. T/F I like to touch and be touched when appropriate.
6. T/F I don't trust most people.
7. T/F I am comfortable when others disclose personal matters to me.
8. T/F If someone compliments me I accept it easily.
9. T/F I really enjoy just doing nothing at all.
10. T/F I get along well with others.
11. T/F It is important that I be right when we disagree.
12. T/F Keeping very busy helps things in my relationships.
13. T/F It's easy for me to understand the emotions and experiences of others.
14. T/F I hate it when my partner feels vulnerable.
15. T/F Being kind is more important than being honest.
16. T/F It is easy for me to talk to members of the opposite sex.
17. T/F My needs and wants are usually met in my relationships.
18. T/F I rarely show my feelings to others.
19. T/F I sometimes pretend that everything is alright when it isn't.
20. T/F I am comfortable with disagreements in my relationships.
21. T/F I am attractive to others.
22. T/F I know my partner has had other relationships. I'm O.K. with that.
23. T/F When my partner needs space I struggle to give it to him or her.
24. T/F I am always pleasant in my relationships.
25. T/F I know my partner sometimes looks at others and it makes me jealous.
26. T/F If my partner truly loved me, he/she would know what I want and need.
27. T/F I enjoy receiving gifts.
28. T/F In order for relationships to work out, I put up with a lot of things I don't like.
29. T/F I fall in love quickly.
30. T/F Love is risky but worth it.
31. T/F We often work through our disagreements rather than ignore them.
32. T/F I hesitate to express my true inner feelings because I might get hurt.
33. T/F I enjoy pleasing my partner but don't go out of my way to do it.
34. T/F If people really knew who I was they would love me.
35. T/F I try to conceal my private life from most people.

Key

1 = F, 2 = F, 3 = F, 4 = T, 5 = T, 6 = F, 7 = T, 8 = T, 9 = T, 10 = T, 11 = F, 12 = F, 13 = T, 14 = F, 15 = F, 16 = T, 17 = T, 18 = F, 19 = F, 20 = T, 21 = T, 22 = T, 23 = F, 24 = F, 25 = F, 26 = F, 27 = T, 28 = F, 29= F, 30 = T, 31 = T, 32 = F, 33 = T, 34 = T, 35 = F

Grade your assessment using this key. Count the number of correct answers and record it here _____. This assessment is keyed toward intimacy. That means the higher your score the higher the quality of intimacy you probably experience. Intimacy is often established as we balance: risks versus closeness, self needs versus others' needs, and disclosure versus privacy. The items you missed above can be considered in terms of these balances or imbalances. Do the findings from this assessment accurately reflect you, your expectations, and your life experience? Why?

Quality Communications Assessment

In this assessment think about your current or most recent relationship. An answer of T (True) indicates that the item applies and F (False) that it does not. Consult the key only after you have completed the assessment.

1. T/F   Disagreements almost never occur in our relationship.
2. T/F   We consider ourselves to be friends.
3. T/F   I tend to keep feelings locked up inside until I can't take it anymore.
4. T/F   I feel free to talk about anything with my partner.
5. T/F   I can trust my partner with my sensitive feelings.
6. T/F   When we argue, one of us almost always ends up with hurt feelings.
7. T/F   Sometimes I say one thing but really mean something else.
8. T/F   We try to be truthful and honest with one another.
9. T/F   We respect each other as individuals.
10. T/F   We often express our love for one another.
11. T/F   It is easy for us to forgive if we offend one another.
12. T/F   If my partner violates my trust, I let him or her know about it.
13. T/F   I am rarely treated fairly by my partner.
14. T/F   We often blame and accuse one another.
15. T/F   There is give and take in our interactions.
16. T/F   I'd rather not disagree or argue no matter what the costs.
17. T/F   When we disagree, we eventually get to the "heart" of the matter.
18. T/F   We usually uncover our wants, needs, and feelings as we interact.
19. T/F   I want my partner to disclose personal information and I allow it to occur.
20. T/F   We don't always agree on issues. That's what I really like about our relationship.
21. T/F   One of us inevitably gets his or her way even at the other's expense.
22. T/F   When we get really mad, we tend to focus on the behavior not the person.
23. T/F   We can go for weeks without talking about our concerns.
24. T/F   We are very competitive and it is important to keep score on who wins each argument.
25. T/F   We watch 4–6 hours of television per day.
26. T/F   Sometimes the threat of violence resolves the disagreement.
27. T/F   We really listen to one another when it's important.
28. T/F   If one of us has a major concern we make time to deal with it.
29. T/F   We never negotiate. It is always about giving in and guilt.
30. T/F   There are things about each other we dislike but have simply learned to live with.

Key

1 = F, 2 = T, 3 = F, 4 = T, 5 = T, 6 = F, 7 = F, 8 = T, 9 = T, 10 = T, 11 = T, 12 = T, 13 = F, 14 = F, 15 = T, 16 = F, 17 = T, 18 = T, 19 = T, 20 = T, 21 = F, 22 = T, 23 = F, 24 = F, 25 = F, 26 = F, 27 = T, 28 = T, 29 = F, 30 = T

This assessment is coded toward healthy and experienced communications patterns. Give yourself 1 point for each correct answer. A higher score indicates a higher quality of communication. Look at the items you missed and discuss them with your partner. Would improvement in these areas lead to improvement in the overall communications in the relationship? Do the findings from this assessment accurately reflect you, your expectations, and your life experience? Why?

How Are You Attracted to Partners?

A nswer the questions below in reference to your current relationship and all of the past significant relationships you can recall. This assessment will clarify for you how you tend to be attracted to your partners.

List the first names of all the significant relationships you want to consider for this assessment:

1.   4.   7.   10.

2.   5.   8.   11.

3.   6.   9.   12.

Initially, what was it about these persons that drew you to them?

1.   4.   7.   10.

2.   5.   8.   11.

3.   6.   9.   12.

After having gone out a few times, what values and traits about the person made you want more?

1.   4.   7.   10.

2.   5.   8.   11.

3.   6.   9.   12.

Why did you decide to take on roles (steady, exclusive dating, living arrangements, etc. . . . .)?

1.   4.   7.   10.

2.   5.   8.   11.

3.   6.   9.   12.

Why did the relationship end?

1.   4.   7.   10.

2.   5.   8.   11.

3.   6.   9.   12.

Look at your answers for each question. Do patterns emerge such as:

1. What tends to initially draw you to your partners (attractions)?

2. What values and traits tend to make you want more with them?

3. What typically motivates you to move into relationship roles?

4. After considering why each relationship ended, think about why initial attractions, later attractions, and commitments into roles did not keep the relationship together. Any trends?

5. Of all the relationships considered above, which had the highest level of attractions? Why? How does your answer to question 5 impact current potential relationships?

Speaking and Hearing Your Love Styles

Self-Version

In this assessment, you will discover how you communicate love and how you hear other's expressions of love toward you. Place a check mark in all that apply to you in both sections below.

Section 1. How You Communicate Love

_____ 1. I often say "I love you."	_____ 16. I give quality time.
_____ 2. I write love notes.	_____ 17. I earn money.
_____ 3. I say things that indicate my love.	_____ 18. I buy gifts.
_____ 4. I share my feelings and thoughts.	_____ 19. I do chores.
_____ 5. I avoid saying hurtful comments.	_____ 20. I make something.
_____ 6. I express approval.	_____ 21. I do things others like to do.
_____ 7. I let others know they are O.K.	_____ 22. I change my schedule if needed.
_____ 8. I often give praise.	_____ 23. I act grateful when receiving gifts.
_____ 9. I look for ways to compliment.	_____ 24. I give money.
_____ 10. I accept others as they are.	_____ 25. I touch non-sexually.
_____ 11. I give of my time.	_____ 26. I hug.
_____ 12. I do regular activities with others.	_____ 27. I kiss.
_____ 13. I slow down to simply talk or listen.	_____ 28. I hold hands.
_____ 14. I arrange quiet time together.	_____ 29. I caress or scratch.
_____ 15. I am around a great deal.	_____ 30. I make love.

Section 2. How You Hear Expressions of Love from Others

_____ 31. I need to hear, "I love you."

_____ 32. I like to receive love notes.

_____ 33. I like indirect indications of love.

_____ 34. I need to hear feelings and thoughts.

_____ 35. I don't want hurtful comments.

_____ 36. I like to hear approval.

_____ 37. I need to know I'm O.K.

_____ 38. I need to be praised.

_____ 39. I like to receive compliments.

_____ 40. I need to be accepted as I am.

_____ 41. I need interpersonal time.

_____ 42. I need regular activities.

_____ 43. I need to simply talk or listen.

_____ 44. I need quiet time together.

_____ 45. I need to be together a lot.

_____ 46. Receiving quality time.

_____ 47. Having financial support from others.

_____ 48. Receiving store-bought gifts.

_____ 49. Having my chores done for me.

_____ 50. Receiving a handmade gift.

_____ 51. Doing things I like to do.

_____ 52. Others adjusting schedules for my needs.

_____ 53. Having my gifts appreciated.

_____ 54. Receiving money.

_____ 55. Non-sexual touching.

_____ 56. Hugging.

_____ 57. Kissing.

_____ 58. Holding hands.

_____ 59. Caressing or scratching.

_____ 60. Making love.

Place the number of the statements (1–60) in the boxes below. This gives you a general idea of how you communicate (1–30) and hear (31–60) love messages. What trends emerge for you?

Verbal	Time Related	Service or Doing	Exchanges	Acceptance	Interactions	Closeness

Speaking and Hearing Your Love Styles

Partner Version

In this assessment, you will discover how you communicate love and how you hear other's expressions of love toward you. Place a check mark in all that apply to you in both sections below.

Section 1. How You Communicate Love

_____ 1. I often say "I love you."	_____ 16. I give quality time.
_____ 2. I write love notes.	_____ 17. I earn money.
_____ 3. I say things that indicate my love.	_____ 18. I buy gifts.
_____ 4. I share my feelings and thoughts.	_____ 19. I do chores.
_____ 5. I avoid saying hurtful comments.	_____ 20. I make something.
_____ 6. I express approval.	_____ 21. I do things others like to do.
_____ 7. I let others know they are O.K.	_____ 22. I change my schedule if needed.
_____ 8. I often give praise.	_____ 23. I act grateful when receiving gifts.
_____ 9. I look for ways to compliment.	_____ 24. I give money.
_____ 10. I accept others as they are.	_____ 25. I touch non-sexually.
_____ 11. I give of my time.	_____ 26. I hug.
_____ 12. I do regular activities with others.	_____ 27. I kiss.
_____ 13. I slow down to simply talk or listen.	_____ 28. I hold hands.
_____ 14. I arrange quiet time together.	_____ 29. I caress or scratch.
_____ 15. I am around a great deal.	_____ 30. I make love.

Section 2. How You Hear Expressions of Love from Others

_____ 31. I need to hear, "I love you."	_____ 46. Receiving quality time.
_____ 32. I like to receive love notes.	_____ 47. Having financial support from others.
_____ 33. I like indirect indications of love.	_____ 48. Receiving store-bought gifts.
_____ 34. I need to hear feelings and thoughts.	_____ 49. Having my chores done for me.
_____ 35. I don't want hurtful comments.	_____ 50. Receiving a handmade gift.
_____ 36. I like to hear approval.	_____ 51. Doing things I like to do.
_____ 37. I need to know I'm O.K.	_____ 52. Others adjusting schedules for my needs.
_____ 38. I need to be praised.	_____ 53. Having my gifts appreciated.
_____ 39. I like to receive compliments.	_____ 54. Receiving money.
_____ 40. I need to be accepted as I am.	_____ 55. Non-sexual touching.
_____ 41. I need interpersonal time.	_____ 56. Hugging.
_____ 42. I need regular activities.	_____ 57. Kissing.
_____ 43. I need to simply talk or listen.	_____ 58. Holding hands.
_____ 44. I need quiet time together.	_____ 59. Caressing or scratching.
_____ 45. I need to be together a lot.	_____ 60. Making love.

Place the number of the statements (1–60) in the boxes below. This gives you a general idea of how you communicate (1–30) and hear (31–60) love messages. What trends emerge for you?

Verbal	Time Related	Service or Doing	Exchanges	Acceptance	Interactions	Closeness

What Type of Lover Are You?

The 35 questions listed below are designed to give you an idea of which of John Lee's love types you currently are. It first appeared in *Psychology Today* in an article titled, "The Styles of Love" in 1973 and 1974. Answer each question while considering your current or most recent relationship. Use one of the following responses for each question: AA = almost always true; U = usually true; R = rarely true; or AN = almost never true. Don't look at the key until you've answered all 35 questions.

Question	Answer
1. You consider your childhood less happy than the average of peers.	_____
2. You were discontented with your life (work, etc.) at the time your relationship began.	_____
3. You have never been in love before this relationship.	_____
4. You want to be in love or have love as a security.	_____
5. You have a clearly defined ideal image of your desired partner.	_____
6. You felt a strong gut attraction to your beloved on your first encounter.	_____
7. You are preoccupied with thoughts about the beloved.	_____
8. You believe your partner's interest is at least as great as yours.	_____
9. You are eager to see your beloved almost every day; this was true from the beginning.	_____
10. You soon believed this could become a permanent relationship.	_____
11. You see "warning signs" of trouble but ignore them.	_____
12. You deliberately restrain frequency of contact with your partner.	_____
13. You restrict discussion of your feelings with the beloved.	_____
14. You restrict display of your feelings with the beloved.	_____
15. You discuss future plans with your beloved.	_____
16. You discuss a wide range of topics/experiences with your beloved.	_____
17. You try to control the relationship but feel you've lost control.	_____
18. You lose the ability to be the first to terminate the relationship.	_____
19. You try to force the beloved to show more feeling/commitment.	_____
20. You analyze the relationship/weigh it in your mind.	_____
21. You believe in the sincerity of your partner.	_____
22. You blame your partner for difficulties in your relationship.	_____
23. You are jealous and possessive but not to the point of angry conflict.	_____

24. You are jealous to the point of conflict, scenes, threats, and so on. ———

25. Tactile, sensual contact is very important to you. ———

26. Sexual intimacy was achieved early/rapidly in the relationship. ———

27. You take the quality of sexual rapport as a test of love. ———

28. You are willing to work out sex problems/improve techniques. ———

29. You have a continued high rate of sex/tactile contact in the relationship. ———

30. You declare your love first, well ahead of your partner. ———

31. You consider your love life your most important activity, even essential. ———

32. You are prepared to "give all" for love once under way. ———

33. You are willing to suffer abuse, even ridicule, from your partner. ———

34. Your relationship is marked by frequent differences of opinion/anxiety. ———

35. The relationship ends with lasting bitterness/trauma for you. ———

Key to John Lee's Love Types (blanks indicate no hit for this love type)

	1	2	3	4	5	6	7	8	9	10	11	12
Eros	R	R		R	AA	AA	AA		AA	AA	R	AN
Ludus				AN	AN	R	AN	U	AN	AN	R	AA
Storge	AN	AN			AN	AN	AN	R	R	R		R
Mania	U	U		AA	AN	R	AA	AN	AA	AN	AA	R
Ludic Eros		R	U		U					R		R
Storgic Eros			R	AN	AN	AN			R	AA	AN	R
Storgic Ludus			AN	AN	R		R	R	AN	AN	R	U
Pragma			R	U	AA			U	R	U	R	

	13	14	15	16	17	18	19	20	21	22	23	24
Eros	R	R	AA	AA	AN	AN	AN		AA	R	U	AN
Ludus	AA	AA	R	R	AN	AN	AN			U	AN	AN
Storge	U	R	R		AN			AN		R	R	AN
Mania	U	U			AA	AA	AA	U	U	U		AA
Ludic Eros	R	R			AN	R			R	R	R	R
Storgic Eros				U	AN	U	AN	R	U	AN	AN	AN
Storgic Ludus	U	U	AN	R		R	R	R	AA			AN
Pragma	U	U	AA	AA		R		AA				AN

	25	26	27	28	29	30	31	32	33	34	35
Eros	AA	AA	AA	U	U		AA	U		R	AN
Ludus			U	R		AN	AN	AN	AN	AA	R
Storge	AN	AN	AN		R	R	R	U	R	R	R
Mania		AN		R	R	AA	AA	AA	AA	AA	AA
Ludic Eros	U	U	U	U	U			R		R	R
Storgic Eros	AN	R	AN		R	AA	AA	AA		R	AN
Storgic Ludus		U	U	R			R	R	R		R
Pragma	R		R	U	R		R	R	AN	R	R

Use this scale for scoring your test: **AA = 3, U = 2, R = 1, and AN = 0.** Place each question's score next to the love type indicated in the key. For example if you put AA for question 7, the key indicates that this is a characteristic of Eros and Mania. So you put a 3 in the Eros and Mania boxes below. After you've done this for all 35 questions, add up your totals in each category.

                Totals

Eros   _____

Ludus   _____

Storge   _____

Mania   _____

Ludic Eros   _____

Storgic Eros _____

Storgic Ludus _____

Pragma   _____

Do the findings from this assessment accurately reflect you, your expectations, and your life experience? Why?

Lee, John A. (1988) "Love Styles" in *The Psychology of Love*; Sternberg, R., & Barnes, M. eds. New Haven, CT: Yale U. Press.

Understanding the Love Type Categories and Their Meanings

John Lee claims that each of us has numerous types of love at any given point in time. Some of us have some or all of them while others have only a few. Here are simple descriptions of each love type:

Eros is the love of beauty, sensuality, and the erotic or sexual expressions between two people. Eros needs immediate attention and does not endure well during separation. Some refer to Eros as "fuse" love because it burns hot, burns bright, and will quickly burn out if that is the basis of the couples relationship. Eros is healthy in combination with other love types and is typically present over the life course.

Ludus is a playful love that often torments one's partner. Love is a game that is best described as catch and release. The ludic lover is immature and is in love with falling in love. Ludic love is often referred to as "jerk" love because of the painful feeling of having been used that partners often feel.

Manic is a madness type of love where the couple loves and hates so often and interchangeably that it appears as though these love types occur simultaneously. This love is often referred to as "roller coaster" love because of its ups and downs and overall rough experience. Couples often experience conflict on the heels of intimacy and vice versa.

Storge *(stor-gay)* is the love between best friends. It is a smooth, easygoing love that develops slowly and endures well. It is referred to as "freeway" love because it is like a freeway where you can travel just about anywhere on smooth road with very few bumps or sharp turns. This love is peaceful and enchanting and is also one of the more common types among younger couples today.

Agape *(ah-ga-pay)* is the love of humanity or of all men and women. This love is not very demanding and is patient and understanding.

Pragma is the practical love we feel. It is based on the practical and logical and is expressed often as qualifications in one's partner (i.e.: looks, wealth, personality, sense of humor, etc. . . .). The other combinations of these love types are self-explanatory.

Lee claimed that for a mutually satisfying relationship, partners need to share similar styles of love. How closely related were the findings between you and your partner? Are the differences significant to you or your partner?
    What course of action is called for based on your findings?

Ideal versus Practical Love Styles

Respond to the items below by placing a check mark in either the "Applies to Me" or "Does Not Apply to Me" box. Check only one box per item. Consult the key after you have completed the assessment.

Applies to Me	Items	Does Not Apply to Me
_____	1. Love strikes suddenly, often without warning.	_____
_____	2. The main cause of divorce today is that people fall out of love.	_____
_____	3. Love is about flowers, gifts, and thoughtful notes.	_____
_____	4. Love sweeps you off of your feet.	_____
_____	5. Making love is magical when you are in love.	_____
_____	6. You have to actually "fall" in love for it to be real.	_____
_____	7. Love at first sight is the healthiest type of love.	_____
_____	8. There is a "special" someone out there for everyone.	_____
_____	9. You can tell if you are in love by the quality of your first kiss.	_____
_____	10. Love is too wonderful to understand.	_____
_____	11. The faster you fall in love the more genuine that love is.	_____
_____	12. Real love prevents problems from occurring.	_____
_____	13. Love can't be controlled, it is too powerful.	_____
_____	14. Love overcomes most differences.	_____
_____	15. True love happens only once in a lifetime.	_____
_____	16. Love is more than holding hands and candlelight dinners.	_____
_____	17. Love makes sense.	_____
_____	18. It helps if people who are in love have other things in common.	_____
_____	19. Love is important but it doesn't solve relationship problems.	_____
_____	20. Healthy relationships develop gradually over time.	_____
_____	21. There is more to a relationship than hugging and cuddling.	_____
_____	22. It is very important to consider who your partner is and what he or she is like before you fall in love.	_____
_____	23. Love is a complex experience but it can be understood.	_____
_____	24. Partners really do need to like and love each other.	_____

_____ 25. Love is a calm thing. _____

_____ 26. Each of us may love more than a few times throughout life. _____

_____ 27. I would marry someone even if I wasn't deeply in love. _____

_____ 28. It takes time to get to know someone before you can fall in love. _____

_____ 29. You have to be a clear-thinking person in order to fall in love. _____

_____ 30. Differences in tastes, hobbies, and career goals can be damaging to a relationship. _____

Key

In items 1–15 count up the number of check marks in the "Applies to Me" and the "Does Not Apply to Me" columns and place that number in the boxes below. Then do the same for items 16–30.

	"Applies to Me"	"Does Not Apply to Me"
Items 1–15		
Item 16–30		

Add up the total numbers in the two *shaded boxes*. This represents your ideal or romantic tendencies. Put your score here: _____ (0–30 possible). The closer this score is to **30** the more ideal and romantic you tend to be in relationships.

Add up the total numbers in the two *clear boxes*. This represents your practical or realistic tendencies in relationships. Put your score here: _____ (0–30 possible). The closer this score is to **30** the more practical or realistic you tend to be in relationships.

Do the findings from this assessment accurately reflect you, your expectations, and your life experience? Why or why not? Did you score about even in both boxes? Many do. Why do we balance ideal and practical?

How Healthy Are Your Break Ups?

Answer the questions T (True) or F (False) below as you think about how your past relationships tend to end. These questions relate specifically to patterns of breaking up. Don't look at the key until you've completed all the questions.

1. T/F   I can usually tell when the relationship is over.
2. T/F   Most of my break ups are not that traumatic.
3. T/F   I typically do most of the breaking up.
4. T/F   I know from the start of a relationship if it will last or not.
5. T/F   If my partner does the breaking up I get even.
6. T/F   I'll do anything to avoid breaking up.
7. T/F   After we break up we try to remain close friends.
8. T/F   Sometimes the break up helps the relationship get better.
9. T/F   Smart people can break up to get more from their partners after they make up.
10. T/F   Breaking up makes the ongoing relationship better.
11. T/F   I will put off a break up until other options present themselves.
12. T/F   In order for me to break up, my partner must not suffer.
13. T/F   Nobody breaks up with me.
14. T/F   Kissing and hugging is good for people who have broken up.
15. T/F   I get along well at a distance when I break up.
16. T/F   Breaking up is the worst event that can happen.
17. T/F   Physical violence has no place in a relationship.
18. T/F   I'll tell my friends the intimate details of a broken up relationship.
19. T/F   If I sense trouble in the relationship, then I will break it off first.
20. T/F   I can date a former partner's close friend or roommate.
21. T/F   I never really know how my relationships end in break ups.
22. T/F   I have to write a break up letter rather than discuss it.
23. T/F   A good break up includes lots of affection.
24. T/F   It's best to have a friend tell your partner you are breaking up.
25. T/F   With some people you have to protect yourself if breaking up.
26. T/F   A break up should be permanent.
27. T/F   A break up should have clearly severed ties.
28. T/F   A break up should be violence free.
29. T/F   Emotional pain is a natural part of the break up.
30. T/F   A break up should not be a punishment.

Key

1 = T, 2 = T, 3 = T, 4 = T, 5 = F, 6 = F, 7 = F, 8 = T, 9 = F, 10 = F, 11 = F, 12 = F, 13 = F, 14 = F, 15 = T, 16 = F, 17 = T, 18 = F, 19 = F, 20 = F, 21 = F, 22 = F, 23 = F, 24 = F, 25 = T, 26 = T, 27 = T, 28 = T, 29 = T, 30 = T

Your score _____

This assessment is biased toward healthy break up patterns. A score of **30** means very healthy patterns, **0** = very unhealthy ones. Breaking up may be a human phenomena that is not considered healthy by nature. But break ups can go well or poorly. Consider some of the items you missed above in terms of your recent break ups. Do they shed any light? Talk to a close friend or family member about any issues that arise from this assessment. List the major concerns you have when break ups occur in your life.

Open and Closed Families

amilies range between being "open" and "closed." Families that are closed and who isolate themselves from society create an environment that could be termed as "psychological incest." In this type of family, children are unable to evaluate their parents in contrast to other families. This occurs especially when parents both agree on their philosophy to their children. Children do not learn or perceive differences and may follow blindly whatever their parents teach them. Idiosyncrasy becomes extreme in isolation.

When families are open, children are able to learn about other families and other lifestyles. They are able to evaluate what they think is appropriate in society. Extended families are important.

Families can be too open when their boundaries are blurred and individuals within the family are always changing and a sense of stability is lost.

On a scale of 1–10, (**1 being isolated, 10 being too open, and 5–6 favorable**), evaluate the openness or closeness of the following.

Your family:

       1   2   3   4   5   6   7   8   9   10

Partner's family:

       1   2   3   4   5   6   7   8   9   10

What you want for your family:

       1   2   3   4   5   6   7   8   9   10

Partner:

       1   2   3   4   5   6   7   8   9   10

Together:

       1   2   3   4   5   6   7   8   9   10

Plans to incorporate:

Boundary Maintenance Questionnaire

In the questionnaire below, think of the relationships you currently have and those you have had most recently with family, friends, and associates. Select T (True) or F (False) for each item. Look at the key only after you have finished the questionnaire.

1. T/F I am honest most of the time in my relationships.
2. T/F If ever I am embarrassed by someone, I let that person know.
3. T/F I am aware of my wants and needs in my relationships.
4. T/F I often communicate my wants and needs to others.
5. T/F I am a person of equal worth to most with whom I interact.
6. T/F If my parents get overly personal I let them know it.
7. T/F I know what I want in life and I pursue it.
8. T/F When I interact, I weigh the needs of others against my own needs.
9. T/F I usually accept compliments when given to me.
10. T/F I greet the challenges of most days with a positive outlook.
11. T/F I enjoy communicating with others.
12. T/F I often assert myself when it is appropriate.
13. T/F I am in tune with my feelings and act accordingly.
14. T/F When something bothers me I talk about it with those involved.
15. T/F People understand me.
16. T/F I know what I want in life.
17. T/F I am comfortable communicating with most people.
18. T/F I am aware of my skills, abilities, and talents.
19. T/F I stop to negotiate when I feel I'm giving too much in the relationship.
20. T/F I am just as valuable as my dates, spouse, partner, or significant other.
21. T/F Sometimes I will focus on the needs of others before mine if the circumstances call for it.
22. T/F I get a lot out of my relationships.
23. T/F I don't let others control and manipulate me.
24. T/F I might be sad if I lost certain relationships but I would eventually get over it.
25. T/F Others appreciate me and value my relationship with them.
26. T/F I let no one touch me in ways I am not comfortable with.
27. T/F Most people who know me respect me.
28. T/F I usually don't pretend that the relationship is going well if it isn't.
29. T/F I will offer criticism at times to others.
30. T/F If someone tells me to "back off" I usually do.

Key

Give yourself 1 point for each true answer, **30** points are possible. Higher scores indicate healthy boundary maintenance in your relationships.

Do the findings from this assessment accurately reflect you, your expectations, and your life experience? Why?

My House, My Boundaries

Understanding Boundaries in Family Systems Theory Using the "My House" Metaphor

The term "boundaries" is a concept used in human relationships and family systems that is basically defined as: "emotional, psychological, or physical separateness between individuals, roles, and subsystems in the family." Social scientists have known for years that boundary maintenance is important for healthy relationships. From the Family Systems perspective we learn that family subsystems need to be maintained properly so that the overall family system functions properly. It is also important that interpersonal boundaries be maintained. But how exactly does one maintain them in our families that tend to be diverse and complex? One answer is to use the house paradigm of personal boundaries. Think of yourself as having a personal house that exists in the suburbs of your many relationships. We put locks and latches on our real house doors and windows to keep intruders out. This paradigm will teach you how to put relationship locks and latches on your personal house so that only those you choose to invite into your house will be allowed in and at a level of interaction that you are comfortable with.

Each of us has the responsibility of taking charge of our own house. That means we choose which people we invite in, when they are invited in, and which level of closeness in our house we allow them to share with us. We also have the responsibility of ensuring that we don't violate others' house boundaries. *(Look at the floor plan on next page.)* Think of it in terms of varying levels of intimate or personal interaction with others. The gate is the most superficial level of interaction; whereas the bedroom is the deepest level. Let's consider each part of the house and its level of intimacy. The **gate** is where we typically interact with strangers. We say "Hello," "Hi," "How's it goin'?" We often don't really want to have the person respond. These are simply polite greetings we use with people we don't know. The **porch and entryway** is for people you are getting to know better, for example, another student you sit next to in class. You might begin to share personal information such as: your name, where you are from, or your major. At these levels you rarely share extremely personal information. That is reserved for people you have known for a long time and already trust.

Let's say that after a few weeks of school, you form a study group including the classmate you previously introduced yourself to. After a few tests and projects, you find that everyone in the study group has been sharing personal information. This might include information about your family, career aspirations, struggles with your parents, and the like. You are now interacting at the **living room** level in your house. You share information but are still guarded about the more vulnerable things about yourself. In the **kitchen** you share more personal information. This you might do with someone you are going steady with, dating, or feel very close to. In this level of interacting, you have deeply established trust and can share your fears, concerns, weaknesses, and hopes with someone in conversation. In the kitchen, confidences are kept. Each knows and respects this fact. The kitchen is often the deepest level of intimacy outside of marriage, cohabitation, or long-term commitments. The **bedroom** represents the level of intimacy that spouses and partners experience. Here a person expresses intimacy at the most intimate level. You can think of the bedroom as representing a haven where physical and emotional intimacy can flourish. In the bedroom we are seen by our partner in our "naked" form. This implies that we are our "true naked self" here. In other words, our spouse or partner accepts us and interacts with us knowing our less-apparent flaws. But even for couples, boundaries must be maintained. Each of us has a **safe.** Our safe represents the most intimate, vulnerable, and personal part of who we are. We rarely open it, even for our spouse or partner. When we do open it we must train our significant other to treat its contents with the utmost respect and dignity. This takes practice, time, and lots of forgiveness for couples to achieve. Only we know the combination to our safe and we choose when to open and close it. You will notice that the bedroom can be attached to another bedroom. You and your spouse or partner each have your own house and each interact in each other's bedroom simultaneously.

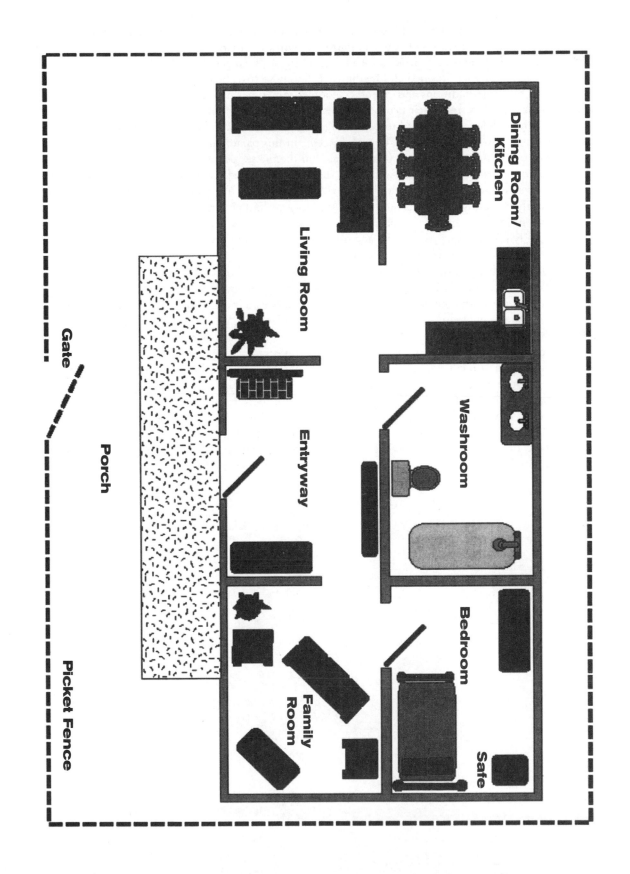

Often newlyweds have the challenge of removing extended family members from the bedroom level (especially parents). We have heard horror stories of parents interfering with their married children's relationship by giving unsolicited financial, sexual, and/or contraceptive advice; setting up financial deals that keep the children indebted to them; and overinvolving their married children in their family so that their children have to struggle to establish their own marriage traditions and customs. Couples sometimes have to be extremely diligent in removing the parents or other offenders from their bedroom issues. If this is the case for you keep in mind that in the long run, it will be worth it. Relationships tend to be healthier and people tend to be happier when boundaries are maintained. The other two rooms represent unique concepts in this paradigm.

The **washroom** represents a place where you can clean up the messes people sometimes bring into your house. For example, sometimes parents get too personal with their newly married children and can be offensive at times. After you and your spouse or partner remove them to the level of interacting you are comfortable with, you can symbolically wash their muddy footprints out of your rug, forgive, and get on with things. The **family room** represents the level of interacting that is appropriate to the family but not necessarily to others outside of the family system. Family jokes, stories, traditions, and other appropriate interactions occur in this room.

As was mentioned before, you are responsible for maintaining the level of interaction in your house. You should interact at levels that you define as comfortable and appropriate for you. Many in our society are conditioned *not* to respect boundaries and most who don't are not even aware of why it is such an unhealthy practice. There are numerous methods you can use to either remove an intruder from rooms in your house or to remove yourself if you are invited into someone else's house at a level you are not comfortable with. First, you might distract the other person by changing the subject or talking to another person in the room. Second, you might confront the person about the level of intimacy he or she is interacting on and why you are uncomfortable with it. Third, you might also consider harshly confronting the other person (the sooner the better in most relationships) or if your previous efforts appear to have brought little change in the relationship.

You are the very best judge of specifically how to maintain your personal boundaries. Keep in mind that this paradigm is based on the belief that personal boundary maintenance is really about interacting with others based on your true feelings, needs, and wants. It is not about controlling others. It is about self-control and to a large degree honesty with our self. We have included a worksheet to assist you in thinking through those strategies you might employ for specific people. We have given you an example of a generic floor plan of the house (you would do best to draw your own).

Boundary Maintenance Worksheet

Sketch the floor plan of your intimacy house and yard or use the one provided for you. Your house should include everything in the list below. Remember, this is *your* house. *You* hold and control all of the keys, latches, and combinations.

Components of House	Level of Intimacy
Fence with a gate	Low
Porch with front door to house	
Entryway/foyer	
Living room	
Dining room/kitchen	Moderate
Family room	
Bathroom/washroom	
Connecting garage with spouse's house (if applicable)	
Bedroom with locking door/windows	
Safe with combination lock in the bedroom	High

Now that you have your house and yard planned out, map out which relationships are appropriate in which place or level. For example, you might complete the phrase:

_____ *(Name of person)* is my _____ *(relationship to*

you). I am comfortable with him/her interacting with me in the _____ *(list room or*

level you feel is appropriate).

_____ *(Name of person)* is my _____ *(relationship to*

you). He/She is often interacting with me in the _____ *(list room or level)* when I do

not want them there. I will remove him/her to the _____ *(list room or level)* by

closing and locking my windows and doors until they are at the level I am comfortable with.

_____ *(Name of person)* is my _____ *(relationship to*

you). He/She often stains or muddies the carpets in my _____ *(list room or level)*. I

will do the following in order to negotiate how I believe they should act in my house:

_____ *(Name of person)* is my _____ *(relationship to you)*. He/She often stays in my _____ *(list room or level)*. I would like him/her to move into the _____ *(list room or level)*. I will do the following in order to invite them to feel comfortable there:

    1.

    2.

    3.

    4.

    5.

What Do You Look for in Your Ideal Mate?

In the items below rank the traits you would like to have in your "ideal mate." Put "A" beside the items that you think are the most important traits. Put "B" beside the items that you desire if possible. Put "C" beside the items you think are not really that important. You may put more than one A, B, or C in each category.

Physical Appearance

_____ Attractive face	_____ Attractive hands	_____ Attractive hair	_____ Attractive figure
_____ Attractive smile	_____ Muscular build	_____ Taller than me	_____ Sexy
_____ Thin–medium build	_____ Medium–large build	_____ Healthy teeth	_____ Nice eyes
_____ Long legs	_____ Fair complexion	_____ Stylish dresser	_____ As attractive as me

Other A's: _____

Personality/Disposition

_____ Moral	_____ Unselfish	_____ Courteous	_____ Honest	_____ Hardworking	_____ Serious
_____ Assertive	_____ Kind	_____ Friendly	_____ Romantic	_____ Acts natural	_____ Interesting
_____ Generous	_____ Social	_____ Creative	_____ Smart	_____ Strong willed	_____ Respectful
_____ Faithful	_____ Playful	_____ Stable	_____ Nice	_____ Confident	_____ Energetic
_____ Committed	_____ Leader	_____ Musical	_____ Warm	_____ Mature	_____ Passionate
_____ Gentle	_____ Candid	_____ Exciting	_____ Clean	_____ Funny	_____ Wild
_____ Open-minded		_____ Appreciative		_____ Comfortable	
_____ Soft-spoken		_____ Independent		_____ Dependable	

Other A's: _____

Communication/Relationship Skills

_____ In tune with own feelings	_____ Shows emotions	_____ Listens well
_____ Understands me	_____ Doesn't blame others	_____ Tells stories
_____ Can handle group situations	_____ Is interesting	_____ Tells jokes
_____ Gives many compliments	_____ Lets me talk too	_____ Can handle disagreements
_____ Exudes confidence	_____ Makes eye contact	_____ Touches and is touchable
_____ Can self-disclose	_____ Doesn't analyze me	_____ Is interested in my life
_____ Accepts me for who I am	_____ Noncompetitive with me	_____ Deals with anger well
_____ Gets along well with parents	_____ Can love and be loved	_____ Sets boundaries well
_____ Knows how to get to the heart of the matter		_____ Can take a compliment

Other A's: _____

Interests

_____ Enjoys family	_____ Enjoys sports	_____ Enjoys music	_____ Enjoys the arts
_____ Enjoys career	_____ Enjoys outdoors	_____ Enjoys parties	_____ Enjoys smoking
_____ Enjoys drinking	_____ Enjoys TV	_____ Enjoys dancing	_____ Enjoys remodeling
_____ Enjoys religion	_____ Enjoys work	_____ Enjoys museums	_____ Enjoys finishing things
_____ Enjoys health	_____ Enjoys sex	_____ Enjoys dining out	_____ Enjoys making money
_____ Enjoys shopping	_____ Enjoys politics	_____ Enjoys planning	_____ Enjoys learning
_____ Enjoys setting goals and attaining them		_____ Enjoys hearing about my day	

Other A's: _____

Family Characteristics

_____ Likes children	_____ Likes marriage	_____ Likes his/her family of origin
_____ Budgets well	_____ Likes being at home	_____ Likes to go out as family
_____ Makes good money	_____ Has parents I like	_____ Balances work/home
_____ Shares with others	_____ Is organized	_____ Shows affection
_____ Comes from wealthy background	_____ Values family traditions	
_____ Maintains long-term relationships	_____ Values education	
_____ Will date me throughout our marriage	_____ Maintains healthy family boundaries	

Other A's: _____

Now take all of your "A" rated ideal traits and put in the box below.

My ideal mate would be someone who . . .

(This list of ideal traits can be read as a paragraph describing your ideal mate.)

Now take your ideal traits and list your top ten most important here (two from each category):

1.   5.   9.

2.   6.   10.

3.   7.

4.   8.

As you date and/or consider serious, long-term commitments, reflect upon these traits you've identified. Does your current partner possess your ideal traits? If not, can you identify what it is that attracts you to him or her? Are you willing to negotiate on some of these traits? Which ones? Have your partner take this and then discuss your ideals.*

Do the findings from this assessment accurately reflect you, your expectations, and your life experience? Why?

---

*Remember that this is an "ideal mate" rating form. Actual results may vary.

My Ideal Wife

This assessment helps you to better understand what you expect a wife to be like. You can do this assessment for yourself or for another person. Answer according to your own understanding and expectations of what the role of wife *should* be like. Consult the instructions below after having considered each item. Mark each item using the following scale: 1 = **unimportant**; 2 = **don't care**; 3 = **important**

Economics

_____ Assertive in the workplace	_____ Go early/stay late gal
_____ Goal-driven	_____ Promotion seeker
_____ Well-paid employee	_____ Expensive tastes in lifestyle
_____ Well-educated	_____ Plays the game
_____ High achiever	_____ Dependable
_____ Extra mile provider	_____ Career oriented

Wife to Husband Relationship

_____ Likes to talk	_____ Passionately skilled in sexual matters
_____ Likes to listen	_____ Fidelity at all times
_____ Shares her day with husband	_____ Is devoted to the relationship
_____ Offers nonsexual verbal affection	_____ Loves husband, flaws and all
_____ Gives back rubs, scratches, and so on	_____ Loyal in public
_____ Supports husband's personal interests	_____ Is willing to do things husband likes

Domestic Skills

_____ Washes dishes and cooks	_____ Is good with the children
_____ Does variety of housework	_____ Does the home/car maintenance
_____ Is not messy	_____ Decorates and arranges
_____ Has reasonable hours	_____ Is committed to extended family relationships
_____ Gets along well with others	_____ Supports husband's friendships
_____ Shares the remote control	_____ Supports husband's career needs

Instructions: First consider all of the items you marked as a "3" or as important. In each of the categories below, pick your top three most important and write the items in the box below.

Economics	Wife to Husband	Domestic
1.	1.	1.
2.	2.	2.
3.	3.	3.

Since this assessment can be done for self or other, think about the current status of things. Are you or the other person up to this standard at this point in time? If not can she now or in the future become more of the person you would like her to be? What tradeoffs exist for you in the wife role? Why must we settle for trade offs? Finally, if change is possible, what is the best way to bring it about? In other words what works and doesn't work in bringing about change for you or her? Discuss these issues with someone who interacts regularly with you and can provide you with some feedback.

My Ideal Husband

This assessment helps you to better understand what you expect a husband to be like. You can do this assessment for yourself or for another person. Answer according to your own understanding and expectations of what the role of husband *should* be like. Consult the instructions below after having considered each item. Mark each item using the following scale: **1 = unimportant; 2 = don't care; 3 = important**

Economics

_____ Assertive in the workplace	_____ Go early/stay late guy
_____ Goal-driven	_____ Promotion seeker
_____ Well-paid employee	_____ Expensive tastes in lifestyle
_____ Well-educated	_____ Plays the game
_____ High achiever	_____ Dependable
_____ Extra mile provider	_____ Career oriented

Husband to Wife Relationship

_____ Likes to talk	_____ Passionately skilled in sexual matters
_____ Likes to listen	_____ Fidelity at all times
_____ Shares his day with wife	_____ Is devoted to the relationship
_____ Offers nonsexual verbal affection	_____ Loves wife, flaws and all
_____ Gives back rubs, scratches, and so on	_____ Loyal in public
_____ Supports wife's personal interests	_____ Is willing to do things wife likes

Domestic Skills

_____ Washes dishes and cooks	_____ Is good with the children
_____ Does variety of housework	_____ Does the home/car maintenance
_____ Is not messy	_____ Decorates and arranges
_____ Has reasonable hours	_____ Is committed to extended family relationships
_____ Gets along well with others	_____ Supports wife's friendships
_____ Shares the remote control	_____ Supports wife's career needs

Instructions: First consider all of the items you marked as a "3" or as important. In each of the categories below, pick your top three most important and write the items in the box below.

Economics	Husband to Wife	Domestic
1.	1.	1.
2.	2.	2.
3.	3.	3.

Since this assessment can be done for self or other, think about the current status of things. Are you or the other person up to this standard at this point in time? If not can he now or in the future become more of the person you would like him to be? What tradeoffs exist for you in the husband role? Why must we settle for trade offs? Finally, if change is possible what is the best way to bring it about? In other words what works and doesn't work in bringing about change for you or him? Discuss these issues with someone who interacts regularly with you and can provide you with some feedback.

Homogamy Self-Assessment

I n this questionnaire, answer each question about yourself, your point of view, or circumstances. Have your partner take the partner version (a close friend or family member may take the partner version if you don't currently have a partner). Don't discuss any of the items with the person taking the partner version until both of you have completed your version.

Race: _____ African American _____ Hispanic _____ Native American _____ Asian _____ Caucasian _____ Other

Marital status: _____ Married _____ Never Married _____ Divorced _____ Widowed _____ Separated

Age in years: _____ years

Years in current relationship: _____ years

Number of times ever divorced: _____ 0 _____ 1 _____ 2 _____ 3 _____ 4+

How many children do you have? _____ 0 _____ 1 _____ 2 _____ 3 _____ 4 _____ 5 _____ 6+

How old is your youngest child? _____ 0–5 _____ 6–12 _____ 13–19 + _____ none

Did you ever cohabit with anyone? _____ Yes _____ No

Were your parents ever divorced? _____ Yes _____ No

Your highest level of education: ___ <High school ___ High school ___ Some college ___ Bachelor degree
___ Graduate degree

Did you suffer from unemployment in the last year? _____ Yes _____ No

How many hours per week do you work for pay? _____ 0 _____ 1–15 _____ 16–29 _____ 30–39 _____ 40+

Estimated total household income: _____ 0–20K _____ 21–40K _____ 41–60K _____ 61–80K _____ 81K+

What is your religious affiliation? _____ or _____ none

Answer the following True/False Questions below:

1. T/F I'm not very involved in the community.
2. T/F I have close friends.
3. T/F I'm not close to my extended family.
4. T/F I moved too often in the past 5 years.
5. T/F I get adequate support from my extended family.
6. T/F I will succeed in life.
7. T/F Overall, I am healthy.
8. T/F Stress sometimes overwhelms me.
9. T/F I am an adequate breadwinner.
10. T/F Religiously, I am not very active.
11. T/F I often donate money to religion.
12. T/F I rarely pray at home.

13. **T/F** I often study sacred writings at home.
14. **T/F** Religion is a great source of strength in our relationship.
15. **T/F** I am satisfied with my level of income.
16. **T/F** Overall, I am satisfied with my life.
17. **T/F** I am not satisfied with my current relationship.
18. **T/F** I am not satisfied as a parent.
19. **T/F** I am satisfied with our sexual relationship.
20. **T/F** My parents aren't happily married.
21. **T/F** My current relationship has changed my life for the better.
22. **T/F** Good relationships are long-term commitments.
23. **T/F** I want this relationship to succeed.
24. **T/F** All things considered, we agree on our aims and goals.
25. **T/F** We often fail to meet each other's needs.
26. **T/F** We have little in common.
27. **T/F** For the most part sex should be avoided before serious commitment.
28. **T/F** We agree when it comes to parenting.
29. **T/F** We experience a great deal of stress from parenting.
30. **T/F** We are equally skilled in parenting.
31. **T/F** We are equally committed to our sex life.
32. **T/F** We tend to pull apart during stress.
33. **T/F** It's not safe to share our feelings in this relationship.
34. **T/F** We trust each other.
35. **T/F** We tend to listen well to each other.
36. **T/F** We communicate equally well.

Homogamy Partner Assessment

In this questionnaire, answer each question about yourself, your point of view, or circumstances. Your partner or another person should have taken the self version. Don't discuss any of the items with the person taking the self version until both of you have completed your version.

Race: _____ African American _____ Hispanic _____ Native American _____ Asian _____ Caucasian _____ Other

Marital status: _____ Married _____ Never Married _____ Divorced _____ Widowed _____ Separated

Age in years: _____ years

Years in current relationship: _____ years

Number of times ever divorced: _____ 0 _____ 1 _____ 2 _____ 3 _____ 4+

How many children do you have? _____ 0 _____ 1 _____ 2 _____ 3 _____ 4 _____ 5 _____ 6+

How old is your youngest child? _____ 0–5 _____ 6–12 _____ 13–19 + _____ none

Did you ever cohabit with anyone? _____ Yes _____ No

Were your parents ever divorced? _____ Yes _____ No

Your highest level of education: ___ <High school ___ High school ___ Some college ___ Bachelor degree
___ Graduate degree

Did you suffer from unemployment in the last year? _____ Yes _____ No

How many hours per week do you work for pay? _____ 0 _____ 1–15 _____ 16–29 _____ 30–39 _____ 40+

Estimated total household income: _____ 0–20K _____ 21–40K _____ 41–60K _____ 61–80K _____ 81K+

What is your religious affiliation? _____ or _____ none

Answer the following True/False Questions below:

1. T/F I'm not very involved in the community.
2. T/F I have close friends.
3. T/F I'm not close to my extended family.
4. T/F I moved too often in the past 5 years.
5. T/F I get adequate support from my extended family.
6. T/F I will succeed in life.
7. T/F Overall, I am healthy.
8. T/F Stress sometimes overwhelms me.
9. T/F I am an adequate breadwinner.
10. T/F Religiously, I am not very active.
11. T/F I often donate money to religion.
12. T/F I rarely pray at home.
13. T/F I often study sacred writings at home.

14. **T/F** Religion is a great source of strength in our relationship.
15. **T/F** I am satisfied with my level of income.
16. **T/F** Overall, I am satisfied with my life.
17. **T/F** I am not satisfied with my current relationship.
18. **T/F** I am not satisfied as a parent.
19. **T/F** I am satisfied with our sexual relationship.
20. **T/F** My parents aren't happily married.
21. **T/F** My current relationship has changed my life for the better.
22. **T/F** Good relationships are long-term commitments.
23. **T/F** I want this relationship to succeed.
24. **T/F** All things considered, we agree on our aims and goals.
25. **T/F** We often fail to meet each other's needs.
26. **T/F** We have little in common.
27. **T/F** For the most part sex should be avoided before serious commitment.
28. **T/F** We agree when it comes to parenting.
29. **T/F** We experience a great deal of stress from parenting.
30. **T/F** We are equally skilled in parenting.
31. **T/F** We are equally committed to our sex life.
32. **T/F** We tend to pull apart during stress.
33. **T/F** It's not safe to share our feelings in this relationship.
34. **T/F** We trust each other.
35. **T/F** We tend to listen well to each other.
36. **T/F** We communicate equally well.

Processing the Homogamy Assessments

This part of the Homogamy assessment requires you to look at the answers to both the Self- and Partner Homogamy Assessments. There are three columns in the following tables. Put the answers beside each item for self and partner. Count up the number of items you both answered the same. If you scored **0** the same, that would imply no homogamy (very unlikely). A score of **50** would indicate perfect homogamy (also very unlikely). Most long-term couples score middle to high homogamy. The score is one indicator, but discussing each item and its importance to you both may be more useful.

Self	Item	Partner	Self	Item	Partner
_____	Race	_____	_____	Rarely pray at home	_____
_____	Marital status	_____	_____	Sacred study at home	_____
_____	Age	_____	_____	Religious source of strength	_____
_____	Years in relationship	_____	_____	Satisfied with income	_____
_____	Times ever divorced	_____	_____	Satisfied with life	_____
_____	How many children	_____	_____	Satisfied with relationship	_____
_____	Age of youngest child	_____	_____	Satisfied as parent	_____
_____	Cohabit	_____	_____	Satisfied with sex	_____
_____	Parents ever divorced	_____	_____	Happily married parents	_____
_____	Level of education	_____	_____	Relationship changed life	_____
_____	Unemployed in last year	_____	_____	Long-term commitments	_____
_____	Work hours per week	_____	_____	Want relationship success	_____
_____	Income level	_____	_____	Agreement on aims/goals	_____
_____	Religion	_____	_____	Fail to meet other's needs	_____
_____	Involved in community	_____	_____	We have little in common	_____
_____	Have close friends	_____	_____	Avoid sex before marriage	_____
_____	Close to extended family	_____	_____	Agree on parenting	_____
_____	Moved too often	_____	_____	Parenting stress	_____
_____	Support from extended family	_____	_____	Equally skilled as parents	_____
_____	Will succeed in life	_____	_____	Equally committed to sex	_____
_____	Healthy overall	_____	_____	Pull apart during stress	_____
_____	Stress overwhelms	_____	_____	Unsafe to share feelings	_____
_____	Adequate breadwinner	_____	_____	Trust each other	_____
_____	Religiously not active	_____	_____	Listen well to each other	_____
_____	Donates money to religion	_____	_____	Communicate equally	_____

Quality of My Dating Experiences

For the following assessment, consider your current or most recent dating experiences. As you consider each item, think about "what is" rather than "what should be or could be." Consult the key only after having completed the assessment. Answer T (True) or F (False). You might consider copying this and having one of your dates take it for comparison.

1. T/F   I tend to date people even if I don't like them that much.
2. T/F   Sometimes my dates get violent.
3. T/F   I get about as much out of dating as my partners.
4. T/F   Dating is fun.
5. T/F   Dating takes my mind off of everyday things.
6. T/F   I enjoy sharing information about myself on the first date.
7. T/F   I often feel overly pressured to kiss, hold hands, or give other forms of physical affection.
8. T/F   I never know what to talk about on a date.
9. T/F   I tend to date only people I would consider for marriage.
10. T/F   The only real way to get close to my date is by having sex.
11. T/F   I usually know what I want to do on a date and express it to my partner.
12. T/F   My partner should do all of the talking on our dates.
13. T/F   I sometimes go out with people I have previously broken up with.
14. T/F   If things get out of control on a date, I'll call it off.
15. T/F   Dating is something people do to find a marriage partner, not just for fun.
16. T/F   I often kiss, hold hands, or give other forms of physical affection when I don't want to.
17. T/F   I dislike doing all of the work on a date.
18. T/F   Dating is about competition for the best-looking and most popular dates.
19. T/F   I have punched/slapped or been punched/slapped on dates.
20. T/F   I tend to have the most power on my dates.
21. T/F   Dating involves taking too many chances.
22. T/F   I have never been able to ask someone out.
23. T/F   When I date "steady" I sometimes will see others secretly.
24. T/F   I just can't tell my date that I don't want to see him or her again.
25. T/F   I typically date people of equal quality.
26. T/F   It's up to my partner to make things go smoothly on the date.
27. T/F   Most of my dates want physical affection and will comply if I pressure them enough.
28. T/F   Rejection is part of the dating game; I am not overly bothered by it.
29. T/F   If it's a good date one of us has to spend money.
30. T/F   If I end up in a social setting that I am uncomfortable with then I'll leave.

Key

Grade your assessment using the following key: 1 = F, 2 = F, 3 = T, 4 = T, 5 = T, 6 = T, 7 = F, 8 = F, 9 = F, 10 = F, 11 = T, 12 = F, 13 = F, 14 = T, 15 = F, 16 = F, 17 = T, 18 = F, 19 = F, 20 = F, 21 = F, 22 = F, 23 = F, 24 = F, 25 = T, 26 = F, 27 = F, 28 = T, 29 = F, 30 = T

This assessment is scored in favor of healthy and meaningful dating practices. Give yourself **1 point** for each correct answer. A higher score indicates higher quality dating. Look at any items you missed and consider why your response might be healthy or unhealthy. Do you disagree with any of these items? How does your date feel about them? What changes do you plan to make if any? Do the findings from this assessment accurately reflect you, your expectations, and your life experience? Why?

Fiancé/Fiancée Communication

Awareness Checklist

e sure to talk about these things with your partner *before* you marry. Chances are you may be surprised by what you learn. If you are not sure about how to start, try using one of the five Ws (who, what, why, when, where, and sometimes how?).

CAUTION: Don't discuss these items with your partner unless you sincerely want to assume the risks of discovering what he or she is truly like.

Money Matters

Who controls the checkbook?

Does your fiancé budget? Do you?

Disposable income:

1. What is a need versus a want?
2. Have things or possessions?
3. Do things or experiences?

Personal cash: How much to each?

Careers

College/Training

1. Who supports whom and when?
2. Whose career takes precedence?

Children

Spacing, who decides?

Discipline styles?

Gender roles?

Father–mother roles?

Birth control?

Allowance, work ethic?

Religious training of children?

Extended Family Relationships

Holidays

Boundaries marriage and extended?

"Ours"; intimate versus everyone's knowledge?

Dealing with differences of opinion?

Household

Cleaning, what and when?

Division of inside and outside tasks?

Child care?

Communication

Argument resolution?

Expression of emotions?

Listening?

Hurt feelings?

Expressing needs and wants?

Intimacy

Techniques?

Self-awareness?

Parent's perspectives on intimacy?

PDA (public display of affection)?

Taboos or "no ways"?

What does intimacy mean to the relationship?

What is intimate and what is public information?

Wedding Plans

What does or doesn't he/she/I want?

Traditions, rituals, honeymoon, and financing?

Marital Anxiety Attitudes Assessment

I n this assessment think about your perceptions of marriage as they currently exist. If married, try to think back on how you felt before you married. Answer T (True) or F (False) on the items below. Consult the key after having completed the assessment.

1. T/F   I hope to marry in the next few years.
2. T/F   Marriage is a very risky thing.
3. T/F   I get scared every time someone close to me divorces.
4. T/F   I don't want to give up dating different people.
5. T/F   I am concerned that after the honeymoon is over, we will not get along so well.
6. T/F   I am concerned that I won't really be able to trust my spouse to be faithful to me.
7. T/F   Getting married means being stuck with the same sexual partner.
8. T/F   Far too many marriages end up in divorce.
9. T/F   I look forward to the deeper levels of intimacy that occur in marriage.
10. T/F   I am afraid that my spouse will get sloppy and unattractive over time.
11. T/F   Most people can't tell what their marriage partner will really be like after the wedding.
12. T/F   My spouse will probably respect me over the years.
13. T/F   I don't want to have arguments when I marry.
14. T/F   I am afraid that someday I might get divorced.
15. T/F   Marriage has too many problems built into it.
16. T/F   Marriage means being stuck with the same person forever.
17. T/F   Marriage is really a constant power struggle.
18. T/F   If I marry, my personal growth will be stifled.
19. T/F   Marriage will enhance my career plans.
20. T/F   In marriage, needs such as love, friendship, and sex are met better than when single.
21. T/F   It doesn't take much to break up a marriage.
22. T/F   Marriage commitments are overwhelming.
23. T/F   Financial support of a marriage and family is extremely costly.
24. T/F   Children are an enormous responsibility.
25. T/F   Married people don't have a lot of fun.
26. T/F   I'll never find my ideal mate.
27. T/F   Many factors beyond our control lead to marital break up.
28. T/F   Marriage means an eventual loss of passion and sensuality.
29. T/F   Marriage facilitates more time shared with the one you love.
30. T/F   Marriage provides emotional security and togetherness.

Key

1 = F, 2 = T, 3 = T, 4 = T, 5 = T, 6 = T, 7 = T, 8 = T, 9 = F, 10 = T, 11 = T, 12 = F, 13 = T, 14 = T, 15 = T, 16 = T, 17 = T, 18 = T, 19 = F, 20 = F, 21 = T, 22 = T, 23 = T, 24 = T, 25 = T, 26 = T, 27 = T, 28 = T, 29 = F, 30 = F

Give yourself **1 point** for each correct answer. This assessment is biased toward marital anxiety. A higher score indicates more anxiety or concern about marriage. Marital anxiety explains, in part, the higher median age at marriage in the United States. Discuss your concerns with your partner, family, or friend. What is their reaction to your concerns? What are their concerns? Do you think your concerns might change over time?

Do the findings from this assessment accurately reflect you, your expectations, and your life experience? Why?

Where You Learned About Sex and Reproduction

Next to each item below, place the letter that corresponds to the appropriate place of origin for your sexual knowledge. Use the following categories: **M** = Mom, **D** = Dad, **OF** = Other Family, **F** = Friends, **TV** = Television, **SC** = School/College, **R** = Religion, **B** = Books I've Read, **P** = Partner(s), or **DK** = Don't Know Where or Did Not Know About It. You may list multiple origins if they apply. Simply place the one that applies the most first.

_____   1. My sexual/reproductive organs

_____   2. My sexual preferences

_____   3. Conception and how it happens

_____   4. Self-pleasuring, stimulation

_____   5. Pregnancy, nine-month gestation, and how it happens

_____   6. Childbirth and how it happens

_____   7. Sexually transmitted diseases

_____   8. Types of contraception

_____   9. The mechanics of sexual expression

_____   10. What uterus, ovaries, and vagina are and how each functions

_____   11. What penis, testicles, and scrotum are and how each functions

_____   12. Why and how breasts develop for women and how breasts function

_____   13. What a monthly period is, and how and why it functions

_____   14. Why sex is "right" sometimes and "wrong" other times

_____   15. My parents' sexual values

_____   16. What a sexual climax is and how it happens

_____   17. How I feel about love, commitment, and sex

_____   18. How to prevent pregnancy

_____   19. How to prevent the spread of diseases

_____   20. That sex on television is not always representative of sex in real-life experiences

_____   21. The social and cultural pressures for or against sex

_____   22. Various sexual techniques and practices

_____   23. What a man is "supposed to do" in bed

_____   24. What a woman is "supposed to do" in bed

_____	25. What menopause is, why and how it happens
_____	26. Why promiscuous people often get a "reputation"
_____	27. Why, with whom, where, when, and how I choose to have sex
_____	28. What sex means to the relationship
_____	29. The role of kissing in sexual expression
_____	30. The role of touch and caress in sexual expression
_____	31. My own sexual boundaries (what I want and don't want)
_____	32. My personal values on sex
_____	33. My values on abortion
_____	34. What virginity means to me
_____	35. What abstinence means to me
_____	36. The risks of premarital sex
_____	37. The occurrence of sexual violence, rape, and abuse
_____	38. The effects of sexual hormones on males and females
_____	39. The processes associated with puberty, dreams, and fantasy
_____	40. The point of view taught in my religion

Once you have completed this assessment, place the item number into the corresponding origin boxes below. If you used more than one origin for an item, place that item in both boxes.

M = Mom	D = Dad	OF = Other Family	F = Friends	TV = Television
Total # of items ____	Total # of items ____	Total # of items ____	Total # of items ____	Total # of items ____
SC = School/College	R = Religion	P = Partner(s)	B = Books I've Read	DK = Don't Know Where or Did Not Know About It
Total # of items ____	Total # of items ____	Total # of items ____	Total # of items ____	Total # of items ____

Answer these insight questions:

1. Rank order the 10 categories above according to the number of items listed in each one. Put the categories with the most items in it first, the second-most second, and so on.

Most Common										

2. After looking at the order of the origins of your sexual knowledge, consider the impact these origins have had on your sexual experience. Would you change their order? If so, why?

3a. Go back to the 40 items. Pick the top 10 that you believe to be the most significant. List their item numbers in the top row of boxes below.

Your Top 10										
Origin Code										

3b. Now take your top 10 picks and locate them on the origin boxes. Put the origin code below the item in the bottom row. Where did the most significant sexual knowledge originate for you? Would you change it if you could? Why or why not? Numerous studies of sexual learning indicate that most men and women learn about sex outside the home. How will you talk to your children about sex? Will you approach it the same way your parents did? Why or why not?

Television Messages About Sexuality

Many studies have established the fact that television viewing shapes our attitudes and outlook on life. Most people in the United States are exposed to numerous television messages while watching 3–4 hours of television per day (that's equivalent to 9.1 years by the age of 65). This project is designed to facilitate an understanding of the sexual messages you get from various television shows. Watch two separate shows from 7–11:00 P.M. and use this table to analyze their presentation of sexuality and sexual issues.

Factors to Consider	Title of First Show	Title of Second Show
Actual count of sexual comments or innuendoes.		
Actual count of sexual acts or acts leading up to sex.		
How many sexual jokes?		
Was sex presented in a relational or nonrelational context?		
Was there any discussion of pregnancy or STD prevention?		
How were sexual scripts and male and female stereotypes presented?		
Were both partners equally interested and involved? How?		
Would you feel comfortable knowing a 10-year-old might be watching these shows? Why or why not?		
Were biased racial and ethnic stereotypes reinforced? How?		
Actual count of sexual messages in commercials.		
Do these television messages about sexuality reflect your own values? Why or why not?		
What was the purpose of sex in the plot of the show?		
Was sex shown as an activity done with alcohol, drug, or other substances?		
Was sex obtained through the use of violence?		
Overall what percentage of the show had sexual themes?		

Childbirth Preferences Questionnaire

I n this questionnaire answer T (True) or F (False). The items are designed to inform you of available childbirth options as well as to clarify your wants, needs, and preferences. Consult the key after completing the questionnaire.

1. **T/F** I prefer to avoid the use of medications during delivery.
2. **T/F** The doctors and nurses should not make any cuts or incisions during delivery.
3. **T/F** The more couples learn about the mother's body and childbirth, the better.
4. **T/F** Relaxation is a better pain manager than drugs.
5. **T/F** Mothers should be as fit as possible for the delivery.
6. **T/F** If circumstances call for it then surgical removal of the baby is acceptable.
7. **T/F** Most women deliver vaginally. It's O.K. to use surgical methods if necessary.
8. **T/F** The mother, father, and doctor should all contribute to the decision to use surgery.
9. **T/F** The health of the baby and mother are the most important considerations.
10. **T/F** Certain women are simply not "built" for vaginal deliveries.
11. **T/F** Couples should take an active role in the delivery.
12. **T/F** The father can be very helpful as a coach for the mother during the delivery.
13. **T/F** A mother can learn to manage delivery pain through breathing and relaxing muscles.
14. **T/F** Delivery is an important event that takes considerable preparation.
15. **T/F** Parents can and should have more control in the delivery process.
16. **T/F** Couples and their new baby are better off if the delivery occurs in the hospital.
17. **T/F** The more training the medical personnel have the better I feel about the delivery.
18. **T/F** Hospitals now have a more "homey" feel about their delivery rooms.
19. **T/F** Hospitals tend to be somewhat flexible in meeting the couple's needs for the delivery.
20. **T/F** Hospitals have all of the medical services the baby and mother could need.
21. **T/F** Home births can be just as safe as hospital births.
22. **T/F** A birth at home is more personal and relaxed than a hospital birth.
23. **T/F** The expenses of a home birth are worth it.
24. **T/F** Most deliveries occur without major medical complications.
25. **T/F** Home births are best attended by qualified medical personnel.
26. **T/F** A spinal block makes the delivery easier for the mother and baby.
27. **T/F** Fetal heart rate monitors are essential for the baby's safety.
28. **T/F** Local anesthetics facilitate an easier delivery for the mother.
29. **T/F** It can be very supporting to have close family and friends in the delivery room at birth.
30. **T/F** Filming or photographing "captures the moments" for the parents, which will be cherished.

Key

These items are designed to make you aware of childbirth options and to give you an opportunity to discover your thoughts and feelings about various options. Items relate to the following childbirth options: **1–5 natural childbirth; 6–10 cesarean section; 11–15 Lamaze; 16–20 hospital birth; 21–25 home birth; and 26–30 miscellaneous but related issues.** Any answer of True indicates an inclination toward these options. Discuss with your partner or other family members why you chose True or False and what your ideal childbirth experience might be.

Death and Grief Assessment

Answer the questions below T (True) or F (False) as they pertain to your life experiences. Please wait until you have completed the assessment before you read the key.

1. **T/F** Death scares me.
2. **T/F** I rarely think about death.
3. **T/F** I am afraid of dying in a painful manner.
4. **T/F** I am troubled about the idea of life, death, and their overall purpose and meaning.
5. **T/F** I rarely go to funerals, even if it is someone close to me who has died.
6. **T/F** I am concerned about a nuclear holocaust.
7. **T/F** We don't talk about death in my family.
8. **T/F** I hate to look at either pictures of dead bodies or real dead bodies.
9. **T/F** When people talk about death, I get real nervous.
10. **T/F** I am concerned with how fast my life appears to be going.
11. **T/F** I am horrified to fly in an airplane because of the possibility of a crash.
12. **T/F** Life is too short.
13. **T/F** I am afraid of contracting AIDS or some other disease.
14. **T/F** Hospital treatments such as operations scare me.
15. **T/F** Hearing about cancer, heart attacks, and strokes makes me feel uneasy.
16. **T/F** In my family, we experience grief and deep feelings of loss when someone dies.
17. **T/F** Grief can be felt in death and in other life events.
18. **T/F** Each person grieves in his or her own way.
19. **T/F** Children are capable of grieving.
20. **T/F** The amount of time required for grieving varies from person to person.
21. **T/F** Keeping busy is not the cure for grief.
22. **T/F** Sometimes people grieve even if death has not occurred.
23. **T/F** It is not very wise to tell a grieving person that you know how he or she feels.
24. **T/F** Listening to a grieving person is a healthy form of support.
25. **T/F** Unresolved childhood grieving may resurface later in life.
26. **T/F** When others grieve, it is helpful to provide support in and around the house.
27. **T/F** Religion and culture strongly influence how we grieve.
28. **T/F** Shock and numbness are normal responses for grievers.
29. **T/F** If you feel sad, it is O.K. to express it to those who are grieving.
30. **T/F** Avoiding the grieving person is not a healthy approach.

Key

Give yourself **1 point** for each True answer in items **1–15**. Also give yourself **1 point** for each False answer in items **16–30**. The highest score you can earn is **30 points**. The lower your score, the healthier your attitude toward death, dying, and grief will be. If you think your score is too high then you might consider taking a course on death and dying or volunteering at a local hospice.

Do the findings from this assessment accurately reflect you, your expectations, and your life experience? Why?

How Strong Is This Relationship?

Answer all of the questions below T (True) or F (False) as you reflect upon your current or most recent relationship. Answer as accurately as possible. Do not read the key until it's completed.

1. T/F   We have close friends whom we associate with.
2. T/F   We aren't close to our extended family.
3. T/F   We've moved too often in the past five years.
4. T/F   We get adequate support from our extended family.
5. T/F   I am not satisfied with our close friendships.
6. T/F   Overall, I am satisfied with our relationship.
7. T/F   My partner and I are best friends.
8. T/F   I like my partner as a person.
9. T/F   We rarely laugh together as a couple.
10. T/F   Overall, we have a happy relationship.
11. T/F   Compared to three years ago, our relationship is worse.
12. T/F   We want each other to succeed in life.
13. T/F   Our friend's relationship is better than ours.
14. T/F   My current relationship has changed my life for the better.
15. T/F   My partner and I are equally happy with our relationship.
16. T/F   My parents aren't happily married.
17. T/F   We rarely eat our main meal together.
18. T/F   We often shop together.
19. T/F   We rarely visit with close friends together.
20. T/F   We rarely work on projects together.
21. T/F   We date each other at a level I am satisfied with.
22. T/F   Our courtship is still going strong.
23. T/F   Our work stress rarely spills over into our relationship.
24. T/F   We have serious quarrels on a regular basis.
25. T/F   We've learned to adjust to each other's habits.
26. T/F   We tend to work out most of our disagreements.
27. T/F   We frequently argue about housework.
28. T/F   I want this relationship to succeed.
29. T/F   Sometimes I think that breaking up would be better than staying together.
30. T/F   All things considered, we agree on our aims and goals.
31. T/F   My partner often fails to meet my needs.
32. T/F   I often fail to meet my partner's needs.
33. T/F   My spouse and I have little in common.

34. T/F   My partner and I agree when it comes to parenting.
35. T/F   I am committed to our sex life.
36. T/F   I am satisfied with our sexual relationship.
37. T/F   We tend to pull apart during stress.
38. T/F   It's not safe to share our feelings together.
39. T/F   We trust each other.
40. T/F   We tend to listen well to each other.
41. T/F   Religiously, I am not very active.
42. T/F   I am satisfied with our level of income.

Key

1 = T, 2 = F, 3 = F, 4 = T, 5 = F, 6 = T, 7 = T, 8 = T, 9 = F, 10 = T, 11 = F, 12 = T, 13 = F, 14 = T, 15 = T, 16 = F, 17 = F, 18 = T, 19 = F, 20 = F, 21 = T, 22 = T, 23 = T, 24 = F, 25 = T, 26 = T, 27 = F, 28 = T, 29 = F, 30 = T, 31 = F, 32 = F, 33 = F, 34 = T, 35 = T, 36 = T, 37 = F, 38 = F, 39 = T, 40 = T, 41 = F, 42 = T

Your Score: _____

Research on strong marriages and relationships has provided insight into the things couples do that facilitate strong relationships. A score of **42** indicates relatively strong patterns, **0** indicates relatively weak ones. Discuss your findings with your partner.

What is the single most important factor contributing to a strong relationship for you?

What is the single most important factor for your partner?

List the three best things your partner does that makes you feel secure in your relationship:

1.

2.

3.

Ask your partner to list three things:

1.

2.

3.

Marital Expectations Assessment

I n this assessment, you will discover your assumptions and expectations about marriage and its impact in your life. Answer each question T (True) or F (False) as accurately as possible. Wait until it is complete before consulting the key.

1. T/F Marriage means happiness.
2. T/F Married people have fewer problems.
3. T/F Married people spend a great deal of time having sex.
4. T/F Married people live more meaningful lives.
5. T/F Married people have it easy.
6. T/F True lovers need not fight.
7. T/F Marriage makes a lot of troubles go away.
8. T/F I can't wait (or couldn't wait) to get married.
9. T/F Love is what gets a couple through the tough times.
10. T/F Our generation has better marriages than did our parent's generation.
11. T/F Parenting simply requires unconditional love.
12. T/F Disagreements can be resolved if the couple focuses on their love.
13. T/F We (will) get along way better than our parents.
14. T/F Cooking, cleaning, and employment skills are not as important as our happiness.
15. T/F Being together is enough to make us happy.
16. T/F Marriage requires some commitment and work.
17. T/F Problems move with singles into their marriage.
18. T/F Sex in marriage is good if the relationship is good.
19. T/F Life is meaningful regardless of one's marital status.
20. T/F Marriage is difficult at times.
21. T/F Disagreeing with a spouse is part of life.
22. T/F Being married gives you a support person for the troubling times.
23. T/F Marriage will come (did come) in a timely manner.
24. T/F When tough times do happen it takes more than love to get through it.
25. T/F All generations of couples have seen both good and bad marriages.
26. T/F Parenting is complex because parents have different ideas about what to do.
27. T/F Disagreements can be resolved if couples focus on the issues leading to the disagreement.
28. T/F We (will) get along at times but not every single minute.
29. T/F Happiness is important but is only part of what marriage is all about.
30. T/F Being apart is just as important as being together.

Key

1–15 = T and represent an extremely idealized expectation of marriage; **16–30** = T but they represent a realistic expectation of marriage. **Your ideal score (questions 1–15)** = _____ and **your realism score (questions 16–30)** = _____. Which score is higher or did you score about the same? Researchers have come to believe that one contributing factor of divorce and marital unhappiness is the overidealized and unrealistically high level of marital expectations. Studies suggest that by starting a marriage with more realistic expectations (or adjusting expectations after the honeymoon has worn off), couples are more personally empowered to negotiate a more stable and happy relationship. Talk about these important issues with someone. What other assessments might help your perceptions of realism?

Do I Have a Strong Family?

Answer the following questions as accurately as possible. You may do this for your family of marriage or origin. Perhaps you could photocopy this quiz, have another family member take it, then discuss your findings.

1. I feel that my family is the best that it can be. Do you agree or disagree (circle one)?

1	2	3	4	5
Strongly Disagree	Disagree	Don't Know	Agree	Strongly Agree

2. I feel that my marriage is the best that it can be. Do you agree or disagree (circle one)?

1	2	3	4	5
Strongly Disagree	Disagree	Don't Know	Agree	Strongly Agree

3. T/F   My family often spends time together having fun.
4. T/F   My family often takes part in traditions.
5. T/F   My family members communicate well. We can talk about most things openly.
6. T/F   My family members express their feelings often.
7. T/F   My family members respect each other's feelings.
8. T/F   My family values working together and we often do.
9. T/F   My family shares the experience of religion and spirituality.
10. T/F   My family members enjoy each other's company.
11. T/F   Each member of my family is valued and important to other members.
12. T/F   When serious problems develop in my family we work them out eventually.
13. T/F   My family allows for each of us to be unique.
14. T/F   Every member of my family is committed to its well-being and success.
15. T/F   We often express appropriate physical affection in my family (hugs, kisses, etc. . . .).
16. T/F   We can turn to one another when things get tough in my family.
17. T/F   We often express verbal affection in my family (love, praise, acceptance, etc. . . .).
18. T/F   We often spend time together just taking it easy.
19. T/F   Each person in my family is allowed to suffer the consequences of his or her own actions.
20. T/F   In my family we have favorite foods that we usually eat on holidays.
21. T/F   In my family, if things get too bad we will seek outside help.
22. T/F   In my family we respect and maintain each other's boundaries.
23. T/F   In my family everyone's input is considered when major decisions are made.
24. T/F   In my family parents are clearly the leaders.
25. T/F   In my family we give each other space when it's needed.
26. T/F   In my family we have healthy relationships with extended family members.
27. T/F   In my family adult children are allowed to make their own decisions.
28. T/F   In my family everyone is supported in his or her interests.
29. T/F   In my family we forgive each other when mistakes are made.
30. T/F   In my family members will listen to each other and are supportive.

Key

Give yourself the number of points that corresponds to the answer you put for questions 1 and 2 (i.e.: if you put 4 for #1 and 2 for #2 you get 6 points). Give yourself **1 point** for each True answer in questions 3–30, then add in your points from questions 1 and 2. If you are not currently married then ignore question 2. The higher your score, the stronger your family is. Strong families are not perfect families. Often families have both strengths and weaknesses. In this quiz you have been exposed to factors found in strong families. If you answered False to some of these questions, talk to a spouse, parent, or other family member about his or her perception of the issue. What if anything could be done to improve this area. Do the findings from this assessment accurately reflect you, your expectations, and your life experience? Why?

An Ideal Family Holiday and Food

In this exercise, pick you favorite holiday of the year. Think about celebrating it over a three-day period. What foods, which people, where, and how does it get celebrated? Also think about the overall function of such a holiday.

1. List the holiday here: _____

2. Plan out the three-day main meal menu:

Day	Who's invited?	Who prepares which part of the meal?	
Day 1 Menu:	Names:	Menu Item:	Name of Preparer:
Day 2 Menu:	Names:	Menu Item:	Name of Preparer:
Day 3 Menu:	Names:	Menu Item:	Name of Preparer:

If the menu listed above is often repeated in family celebrations, what function does it play in your family?

Ultimately, what is it you want to have happen at this holiday celebration?

How does food contribute to that purpose?

Recreation Assessment

In this assessment you will discover how you and your partner enjoy your recreational experiences. Place a check mark in the box if the statement is true for you and do the same for your current or recent partner. Think about recent recreational experiences as you answer these questions. This assessment can be done for one or two people.

Your Name:	Type of Recreational Experience	Your Partner's Name:
	1. Enjoy recreation with large crowds	
	2. Enjoy programmed recreation (amusement and theme parks)	
	3. Enjoy recreation that typically costs money	
	4. Enjoy recreation available near large cities	
	5. Enjoy recreation that is technologically based	
	6. Enjoy recreation that includes nature	
	7. Enjoy recreation that is secluded	
	8. Enjoy recreation that is rural	
	9. Enjoy recreation that has very few people around	
	10. Enjoy recreation that is relatively inexpensive	
	11. Enjoy carefully planned activities	
	12. Enjoy spontaneous, "make it up as you go" activities	
	13. Enjoy overnight trips	
	14. Enjoy road trips	
	15. Enjoy long trips	
	16. Enjoy short getaways	
	17. Enjoy water activities	
	18. Enjoy activities that require exertion (burning calories)	
	19. Enjoy activities where I can meditate lazily	
	20. Enjoy activities where I can interact with people	

Compare and contrast the types of recreation you and your partner prefer. Do themes emerge? Discuss them with your partner. Do you share common preferences? What are your most significant preferences; what are your partner's? Is it possible for each of you to adapt a preference from the other's preference?

Who Has Possession of the Remote Control?

Self-Version

This assessment is designed to clarify the distribution of power in your current or past relationship. Answer each question about yourself and your current or recent partner. Think carefully about each question. In this assessment you answer for yourself and from your perception about your partner. Place a check mark in either your side or your partner's side depending on who it applies to most. If it's about equal then place a check mark in both sides.

Your Name:	Type of Recreational Experience	Your Partner's Name:
	Typically holds the remote control while watching television	
	Typically picks out a video for us to watch	
	Typically chooses which restaurant we will dine at	
	Typically chooses where we go for short getaways	
	Typically chooses the activities for our dates	
	Typically chooses how we spend extra money	
	Typically decides what is a need to be purchased	
	Typically decides what is a want to be purchased	
	Typically chooses the movie we watch at the theater	
	Typically chooses most of the groceries we buy	
	Typically decides our vacation spots	
	Typically controls the checkbook	
	Typically decides which relatives we visit on holidays	
	Typically chooses the radio station we listen to together	
	Typically chooses the CDs we buy	
	Typically dominates the computer	
	Typically spends more time on the Internet	
	Typically takes the pictures with our camera	
	Typically decides when we will be physically affectionate	
	Typically decides how our physical affection is expressed	
	← **Add up the check marks for you and your partner in these boxes** →	

Who has the most check marks, you are your partner? Which marks are in the areas that cause you concern? Discuss this and your partner's assessment and perceptions with your partner. Negotiate changes where needed. Balance does not always imply perfect equality, but in successful relationships balance implies the mutual meeting of both partners needs and wants.

Who Has Possession of the Remote Control?

Partner Version

This assessment is designed to clarify the distribution of power in your current relationship. Answer each question about yourself and your current partner. Think carefully about each question. In this assessment you answer for yourself and from your perception about your partner. Place a check mark in either your side or your partner's side depending on who it applies to most. If it's about equal then place a check mark in both sides.

Your Name:	Type of Recreational Experience	Your Partner's Name:
	Typically holds the remote control while watching television	
	Typically picks out a video for us to watch	
	Typically chooses which restaurant we will dine at	
	Typically chooses where we go for short getaways	
	Typically chooses the activities for our dates	
	Typically chooses how we spend extra money	
	Typically decides what is a need to be purchased	
	Typically decides what is a want to be purchased	
	Typically chooses the movie we watch at the theater	
	Typically chooses most of the groceries we buy	
	Typically decides our vacation spots	
	Typically controls the checkbook	
	Typically decides which relatives we visit on holidays	
	Typically chooses the radio station we listen to together	
	Typically chooses the CDs we buy	
	Typically dominates the computer	
	Typically spends more time on the Internet	
	Typically takes the pictures with our camera	
	Typically decides when we will be physically affectionate	
	Typically decides how our physical affection is expressed	
	← Add up the check marks for you and your partner in these boxes →	

Who has the most check marks, you are your partner? Which marks are in the areas that cause you concern? Discuss this and your partner's assessment and perceptions with your partner. Negotiate changes where needed. Balance does not always imply perfect equality, but in successful relationships balance implies the mutual meeting of both partners needs and wants.

My Ideal Obituary/Life Goals

I n the space provided below, write an obituary for yourself using the following criteria: Assume that it was written in your 85th year (not at your current age); include the personal things you hope will be said about you; and include your ideals and aspirations or things you hope to have accomplished by age 85 (150 words or less). *It might be helpful for you to read a few real obituaries in order to get an understanding of what to include.*

My Obituary at Age 85

Now that your ideal obituary has been written, extract those ideal personal traits and ideal accomplishments that you identified and use them in the assignment below. List the personal traits that you do not now possess and set goals and strategies for achieving these traits in the future. Do the same thing for the ideal accomplishments. Think in terms of what you will have to start, change, or do in order to become the type person you described in your ideal obituary.

Traits and Accomplishments	Strategies and Goals			
	1 Year	5 Year	10 Year	20 Year
Personal Trait 1				
Personal Trait 2				
Accomplishment 1				
Accomplishment 2				
Accomplishment 3				

How Has Parenthood Changed You?

Most researchers have expended considerable resources in the study of the effects of parents on children. Very few studies in comparison have studied the effect of children (parenting) on parents. This exercise will allow you as a parent to observe some of the changes that have taken place in your life. First try to remember how things were before you became a parent. Once you have completed the Before section, go to the After Child(ren) section and complete it. Next, briefly summarize your evaluation of these changes.

Area of Your Life	Before Child(ren)	After Child(ren)	Your Evaluation
Overall Physical Health			
Figure/Build			
Fitness			
Frequency of Illnesses			
Risks of Accidents			
Daily Nutrition			
Overall Happiness			
Family Satisfaction			
Career Satisfaction			
Marital Satisfaction			
Social Life Satisfaction			
Spiritual Satisfaction			
Educational Satisfaction			
Intimacy Satisfaction			
Type of Vehicle You Desire			
Type of Recreation You Desire			
Daily Levels of Stress			
Type of Music You Prefer			
How You Unwind or Relax			
Type of Employment That Fits Your Needs			
Relationship With Parents			

Area of Your Life	Before Child(ren)	After Child(ren)	Your Evaluation
Relationship with In-Laws			
Connection to Community			
Connection to Religion			
Overall Life Plans			
Feelings of Control of Your Life's Course			
Feeling of Adequacy as a Parent			
Conservatism			
Liberalism			
Intellectual Interests			
Casual Reading			

Consider These Questions for This Exercise

How many of the changes you observed would have occurred even if you had no children?

Overall what is your evaluation of the effects that children have had on your life?

If some of the areas indicated no change, do you think that change may yet occur? If so how do you feel about it?

How will life be after your child(ren) are gone? What types of adjustment will be necessary for that transition in life?

Notes:

Single Parenting Assessment

This assessment is specifically designed for single parents. Other parents can take it because many of the parenting principles are similar regardless of marital status. Also, if you are not a parent, it might be very insightful for you to interview a single parent using this questionnaire. Either way, wait until the assessment is completed before looking at the key. *The use of the word,* children *in the questions is meant to imply one or more children.*

1. T/F I have a good life.
2. T/F My emotional needs are meet.
3. T/F I could use more time spent in the company of adults.
4. T/F I have good friendships.
5. T/F I exercise regularly.
6. T/F I get adequate sleep at night.
7. T/F I manage stress successfully.
8. T/F I am not overweight.
9. T/F I look attractive.
10. T/F I have a nice wardrobe.
11. T/F I enjoy the place we live in.
12. T/F I have my intimacy needs met.
13. T/F I have enough money.
14. T/F I have adequate privacy.
15. T/F I pursue my hobbies.
16. T/F I rarely go to bed tired.
17. T/F My social life works for me.
18. T/F I enjoy my extended family.
19. T/F I enjoy parenting.
20. T/F My children are basically unhappy.
21. T/F My children do not listen to me.
22. T/F My children get more of what they want from me than I would like.
23. T/F My children need another parent.
24. T/F My children need more of my time.
25. T/F My children take me for granted.
26. T/F My children are not fed to my standards.
27. T/F My children run over me sometimes.
28. T/F My children drain my energy.
29. T/F My children are not learning responsibilities.
30. T/F My children spend too much time without me.

Key

1–19 = True; 20–30 = False

Your Score _____

This assessment is biased toward healthy single parenting. A score of **30** means very healthy and **0** means very unhealthy single parenting. A single parent often has more responsibilities than couple parents. The first 19 questions deal with the single parent's well-being as an individual. Questions 20–30 deal with the single parent's perception of his or her parenting performance by measuring outcomes of child well-being. Discuss the issues presented in each item. Which is/are the most critical? Which needs immediate attention? What can be done, given the single parent's resources to bring about change?

Fatherhood Patterns Assessment

T his assessment is designed to explore your impression of fatherhood. By answering the questions below, you will become familiar with the types of patterns exhibited in fathering behaviors, especially those you saw from your perspective as a child growing into adulthood. As you answer the T (True) and F (False) questions below, think about your own father, stepfather, grandfather, or adult male (only one role if possible) whom you would consider the primary father figure in your growing up stage of life.

1. T/F   With him I could clearly tell when I had done wrong.
2. T/F   The most common form of punishment was physical in nature.
3. T/F   He would forgive my mistakes.
4. T/F   He was never around to know if I was doing something I shouldn't do.
5. T/F   He would forget about what I had done so I could move on.
6. T/F   He punished me with reasonable punishments.
7. T/F   At times, I really felt afraid of him.
8. T/F   We always did things together.
9. T/F   He was frequently at home in the evenings and/or weekends.
10. T/F   He never really knew what I was going through.
11. T/F   He knew who my friends were.
12. T/F   He was a good provider for me.
13. T/F   I felt protected by him.
14. T/F   If people knew what I knew about him, they'd say he was a good father.
15. T/F   I couldn't really talk to him.
16. T/F   He was there to be a part of my life events.
17. T/F   I was satisfied with his involvement with me.
18. T/F   I could wake him up to talk if I needed to.
19. T/F   He would take me shopping with him.
20. T/F   He and I played together regularly.
21. T/F   He touched me in ways I needed.
22. T/F   He was cool around my friends.
23. T/F   More than once, he pulled me out of a jam.
24. T/F   He spent his money on me.
25. T/F   I did not learn much from him.
26. T/F   We worked together on things.
27. T/F   We get along well now.
28. T/F   When I was hurt, he took care of me.
29. T/F   He's the kind of person I'd like to be.
30. T/F   If I were like him, I'd make a good parent.

Key

1 = T, 2 = F, 3 = T, 4 = F, 5 = T, 6 = T, 7 = F, 8 = T, 9 = T, 10 = F, 11 = T, 12 = T, 13 = T, 14 = T, 15 = F, 16 = T, 17 = T, 18 = T, 19 = T, 20 = T, 21 = T, 22 = T, 23 = T, 24 = T, 25 = F, 26 = T, 27 = T, 28 = T, 29 = T, 30 = T

Your Score = _____

This assessment is biased toward the ideal role of fatherhood. A score of **30** represents a very healthy and positive fatherhood. A score of **0** represents a very unhealthy and negative one. Can you identify the expressive and instrumental traits in your father figure? Many people make strong reference to their memories of their father figure when they themselves become adults and parents. How much of this process do you see in your own life? Prior to taking this assessment had your impression of your father figure changed significantly and if so, why? Has it changed after taking this assessment, if so why? Could you talk about this with him? Why or why not?

Motherhood Patterns Assessment

This assessment is designed to explore your impression of motherhood. By answering the questions below, you will become familiar with the types of patterns exhibited in mothering behaviors, especially those you saw from your perspective as a child growing into adulthood. As you answer the T (True) and F (False) questions below, think about your own mother, stepmother, grandmother, or adult female (only one role if possible) who you would consider the primary mother figure in your growing up stage of life.

1. T/F   With her I could clearly tell when I had done wrong.
2. T/F   The most common form of punishment was physical in nature.
3. T/F   She would forgive my mistakes.
4. T/F   She was never around to know if I was doing something I shouldn't do.
5. T/F   She would forget about what I had done so I could move on.
6. T/F   She punished me with reasonable punishments.
7. T/F   At times, I really felt afraid of her.
8. T/F   We always did things together.
9. T/F   She was frequently at home in the evenings and/or weekends.
10. T/F   She never really knew what I was going through.
11. T/F   She knew who my friends were.
12. T/F   She was a good provider for me.
13. T/F   I felt protected by her.
14. T/F   If people knew what I knew about her, they'd say she was a good mother.
15. T/F   I couldn't really talk to her.
16. T/F   She was there to be a part of my life events.
17. T/F   I was satisfied with her involvement with me.
18. T/F   I could wake her up to talk if I needed to.
19. T/F   She would take me shopping with her.
20. T/F   She and I played together regularly.
21. T/F   She touched me in ways I needed.
22. T/F   She was cool around my friends.
23. T/F   More than once, she pulled me out of a jam.
24. T/F   She spent her money on me.
25. T/F   I did not learn much from her.
26. T/F   We worked together on things.
27. T/F   We get along well now.
28. T/F   When I was hurt, she took care of me.
29. T/F   She's the kind of person I'd like to be.
30. T/F   If I were like her, I'd make a good parent.

Key

1 = T, 2 = F, 3 = T, 4 = F, 5 = T, 6 = T, 7 = F, 8 = T, 9 = T, 10 = F, 11 = T, 12 = T, 13 = T, 14 = T, 15 = F, 16 = T, 17 = T, 18 = T, 19 = T, 20 = T, 21 = T, 22 = T, 23 = T, 24 = T, 25 = F, 26 = T, 27 = T, 28 = T, 29 = T, 30 = T

Your Score = _____

This assessment is biased toward the ideal role of motherhood. A score of **30** represents a very healthy and positive motherhood. A score of **0** represents a very unhealthy and negative one. Can you identify the expressive and instrumental traits in your mother figure? Many people make strong reference to their memories of their mother figure when they themselves become adults and parents. How much of this process do you see in your own life? Prior to taking this assessment had your impression of your mother figure changed significantly and if so, why? Has it changed after taking this assessment, if so why? Could you talk about this with her? Why or why not?

Processing the Fatherhood and Motherhood Patterns Assessments

Now that you have completed the Fatherhood and Motherhood Patterns assessments, you can study the influence these patterns might have or are having in your current roles and relationships. Answer the questions below and discuss them with your partner, parents, or a close friend. Keep in mind that the primary function of these assessments is to help you better understand your perspective and dispositions. It's best to focus more on what you learn rather than how you might stack up in comparison to another person.

Questions

1. Sometimes, in our parenting roles we act more like one parent than the other. Sometimes we act like a mixture of both parents. Which is the case for you? Why is it so?

2. What influence did your parents' other roles have on how they parented? For example, sole versus dual breadwinners, traditional versus working-for-pay mothers, and any other significant adult-family roles.

Compare the father to the mother findings in the assessments (for questions 3, 4, and 5).

3. Who was the most expressive, nurturing, emotionally supportive, and connected with you?

4. Who was the most instrumental, providing, protecting, and economically sustaining for you?

5. Do you anticipate that you will be in similar roles with your partner when you parent?

6. How does your partner compare to the same sex parent you had?

7. How have past partners compared to the same-sex parent you had?

8. Have you discovered any insightful trends or patterns that may be helpful to you in your perception of parent? If so, list them:

9. Why would or wouldn't you talk to your parents about these findings? How does your answer impact your role as parent to your own children?

Children and Attachment

According to studies of childhood attachment, it is important for infants and young children to closely attach to their parents. Early attachment by the young child has been shown to have a positive affect when the child goes through adolescence because of the positive relationships they have with their parents.

During the first few years of life for the infant and young child it is also important for the father and mother to bond and attach to their child.

Some disruptions may occur that affect the child's relationship with their parents. Evaluate how you could counteract some of them in your own experience.

Affects on Attachment of Children to Parents

1. Working parents:

2. Age of child:

       Birth to 5:

3. Extended family:

4. Possibility of mother/father staying home during early years of a child's life and start work when children start school.

5. Possibility of working part-time or sharing flex-time with spouse.

If the mother stays home with the children, it is important for her to stay tied to the workforce by keeping her work skills current. In this way, if she wants or needs to return to work, she can. What would be my plan?

If a woman lives to age 77 and she takes 6–7 years to stay home with infant and toddlers she has given up 7 years out of 77. Is this an option? What is your option?

Assessing the Use of Blame
in Your Relationships

I n this assessment you will become more aware of the use of blame by you against others and by others against you. Answer each question T (True) or F (False) accurately and do not look at the key until you have completed the assessment.

1. T/F   People ought to feel bad for what they do to me.
2. T/F   When things don't work out it's usually somebody else's fault.
3. T/F   Once the other person is aware of their wrongdoings, our relationship goes more smoothly.
4. T/F   It's usually the other person's fault when they don't get what they want.
5. T/F   It isn't easy being the victim at times.
6. T/F   I'm simply better at relationships than most others.
7. T/F   I have to use anger to keep things in control.
8. T/F   I feel a bit vulnerable before arguments.
9. T/F   People have a way of messing things up in my life.
10. T/F   Some of my closest friends are ignorant of what they do to me.
11. T/F   I always have to help people to know how to relate to me.
12. T/F   I struggle with my sense of self-concept.
13. T/F   If you keep at certain people for long enough, they'll figure out what they're doing wrong.
14. T/F   I hate it when people make me look bad.
15. T/F   I see people's behavior so clearly most of the time, but they don't see mine so clearly.
16. T/F   It's always my fault.
17. T/F   I feel bad around certain people but am not sure why.
18. T/F   Unless I did something wrong, others feel uncomfortable around me.
19. T/F   I can never figure out the right thing to do with certain relationships.
20. T/F   I'm often the offender.
21. T/F   Sometimes I feel like not even trying in my relationships.
22. T/F   People often get mad at me and I can't avoid it.
23. T/F   I often blow it for people.
24. T/F   My friends let me know when I mess things up with them.
25. T/F   My needs and wants often cause complications.
26. T/F   Others know I have a low self-concept.
27. T/F   It takes me a long time to figure out people.
28. T/F   I just want us all to be happy.
29. T/F   When I give in we get along better.
30. T/F   I often embarrass my friends and family.

Key

1–30 = True

Your Score _____

This assessment is biased toward the negative and unhealthy use of blame and victimhood. A higher score indicates a greater propensity to use blame or to be the victim of blame. Questions 1–15 relate to patterns of blaming others. Questions 16–30 relate to patterns of being the victim (a convenient facilitator of blaming). How many questions did you score correctly in questions 1–15 ___ and in questions 16–30 ___? This will help you identify if you blame more, are a victim more, or are about equally high or low in these traits. Discuss this with someone you relate to often. Do they see the same thing in you?

Marital Violence Risks Assessment

This assessment can be used by marrieds and singles. Please check all of the factors below that apply to you and your spouse or current partner (simply check spouse if single and it applies to your partner). You may also use it for current or past relationships.

1. I/My spouse was slapped as a child. _____ me _____ spouse

2. I/My spouse was kicked as a child. _____ me _____ spouse

3. I/My spouse was pulled/pushed/shoved as a child. _____ me _____ spouse

4. I/My spouse was bitten/pinched/scratched as a child. _____ me _____ spouse

5. I/My spouse was choked/beaten/threatened with a weapon as a child. _____ me _____ spouse

6. I/My spouse was injured with a weapon as a child. _____ me _____ spouse

7. I/My spouse was ridiculed/cut down/yelled at as a child. _____ me _____ spouse

8. I/My spouse often felt that I/he/she "deserved" what happened in items 1–7. _____ me _____ spouse

9. I/My spouse come(s) from a family where violence is used when it is "needed." _____ me _____ spouse

10. I/My spouse live(s) in a society where violence is used when it is "needed." _____ me _____ spouse

11. I/My spouse break(s) things when angry. _____ me _____ spouse

12. I/My spouse hurt(s) people when angry. _____ me _____ spouse

13. I/My spouse yell(s) harshly when angry. _____ me _____ spouse

14. I/My spouse am/is hurt by others when angry. _____ me _____ spouse

15. I/My spouse sometimes need(s) the help of the police to resolve disagreements. _____ me _____ spouse

16. I/My spouse sometimes find(s) that violence helps to resolve disagreements. _____ me _____ spouse

17. I/My spouse get(s) drunk often. _____ me _____ spouse

18. I/My spouse get(s) high often. _____ me _____ spouse

19. I/My spouse sometimes insult(s) my spouse/me when we are alone. _____ me _____ spouse

20. I/My spouse sometimes insult(s) my spouse/me when we are in public. _____ me _____ spouse

21. I/My spouse get(s) frustrated beyond control because of work or something else. _____ me _____ spouse

22. I/My spouse will not do what I/he/she is "supposed" to do. _____ me _____ spouse

23. I/My spouse will not listen to reason. _____ me _____ spouse

24. I/My spouse get(s) very angry if dinner/the house/or children are "out of order." _____ me _____ spouse

25. I/My spouse feel(s) deeply inadequate as a man/woman. _____ me _____ spouse

26. I/My spouse would physically hurt my spouse/me if he/she had an affair. _____ me _____ spouse

27. I/My spouse would kill my spouse/me if he/she had an affair. _____ me _____ spouse

28. I/My spouse make(s) my spouse/me very angry when I/he/she physically hurt(s) our child. _____ me _____ spouse

29. I/My spouse sometimes scream(s) hysterically. _____ me _____ spouse

30. I/My spouse really feel(s) bad after having physically hurt one another. _____ me _____ spouse

31. I/My spouse can be really mean. _____ me _____ spouse

32. I/My spouse have/has low self-esteem. _____ me _____ spouse

33. I/My spouse am/is not very smart. _____ me _____ spouse

34. I/My spouse have/has a marriage with much conflict. _____ me _____ spouse

35. I/My spouse have/has children with behavioral problems. _____ me _____ spouse

36. I/My spouse stress/es over being a single parent. _____ me _____ spouse

37. I/My spouse frequently lose(s) my/his/her job. _____ me _____ spouse

38. I/My spouse like(s) to keep to ourselves more than socialize. _____ me _____ spouse

39. I/My spouse did not really get along well with friends while growing up. _____ me _____ spouse

40. I/My spouse do(es) not really get along with friends now. _____ me _____ spouse

41. I/My spouse feel(s) that "hard" spankings are sometimes best for the child. _____ me _____ spouse

42. I/My spouse really feel(s) that we "own" our children. _____ me _____ spouse

43. I/My spouse fail(s) to support each other equally in parenting responsibilities. _____ me _____ spouse

44. I/My spouse often have/has many stressful events that appear to "derail" the family. _____ me _____ spouse

45. I/My spouse have/has no use for religion. _____ me _____ spouse

46. I/My spouse, as a parent, rarely receive(s) support from family or friends. _____ me _____ spouse

47. I/My spouse tend(s) to be aggressive toward most people. _____ me _____ spouse

48. I/My spouse use(s) violence only when its really needed. _____ me _____ spouse

49. I/My spouse feel(s) powerful when violence is used. _____ me _____ spouse

50. I/My spouse find(s) that children/spouses/old people/others are difficult to control. _____ me _____ spouse

51. I/My spouse find(s) that when someone gets hurt, they usually "deserve" it. _____ me _____ spouse

52. I/My spouse find(s) that no matter what I/he/she tries, violence tends to occur. _____ me _____ spouse

53. I/My spouse find(s) it very difficult to work things out. _____ me _____ spouse

54. I/My spouse know(s) that violence is unusual. _____ me _____ spouse

55. I/My spouse feel(s) important or valuable as the victim of violence. _____ me _____ spouse

56. I/My spouse feel(s) that I/he/she can change a violent person given enough time. _____ me _____ spouse

57. I/My spouse feel(s) loved and accepted especially after we make up. _____ me _____ spouse

58. I/My spouse believe(s) that violence is caused by genetics. _____ me _____ spouse

59. I/My spouse believe(s) that talking about our violence is inappropriate. _____ me _____ spouse

60. I/My spouse can solve our own problems without outside help. _____ me _____ spouse

The first step in understanding the results of this assessment is to identify which risks are associated with you, your spouse, or both. Place the number of the risk factor in the appropriate box.

Applies to Me	Applies to My Spouse/Partner	Applies to Both of Us

The more items you checked off, the higher the risk that you will end up in a situation of violence in some capacity. It is a well-documented, scientific fact that violent and abusive homes often create violent and abusive people who in turn are often violent abusers or victims of abusers as adults. Two very important principles may be helpful if this is the case for you. First, if you are single, you may want to evaluate carefully your current or most recent significant relationship. Children who were abused often grow up and become abusers themselves or become victims of abusers. Second, if you are married, you may want to consider the fact that violence is rarely overcome in the home without some form of outside intervention. One of the myths of victims and abusers is that it will someday get better if "we just leave it alone."

With your partner or spouse (or on your own if you did this assessment for yourself), seriously consider these risk factors you identified and their repercussions for you, your family, and children. Almost everyone has some propensity toward violence but some of us have more than others. If you and your partner are willing to talk about these risk factors openly and in the context of hope for change you will be able to judge for yourself how best to successfully overcome any problems that currently exist, and to plan how to prevent problems from developing in the future. Talking about it openly and calmly is the first step. Discussing it with appropriate family members, seeking counseling in religious or professional settings, and involving the state in legal interventions are other steps that can be taken when needed.

Rape Awareness Quiz

Answer each question below T (True) or F (False). Read each question in order as you come to it. Don't read the key until you've finished.

1. T/F "Nice" "Good" girls/women don't get raped.
2. T/F Women are raped only when they are in the "wrong" place.
3. T/F Provocative clothing provokes a rapist to attack.
4. T/F Women secretly want to be raped.
5. T/F It is easy to pick a rapist out of a crowd.
6. T/F Rapists are typically strangers to the victim.
7. T/F Rapists are sexually unfulfilled.
8. T/F Rapists come from a different racial or ethnic group than the victim.
9. T/F Rapists usually rape once and only on an impulse.
10. T/F Rape is simply "aggressive sex."
11. T/F Rape is simply "morning after regrets" from willing women.
12. T/F If a rapist does not injure or use violence then it really isn't rape.
13. T/F If a woman does not fight back while resisting a rape attack then it's not rape.
14. T/F It really is the victim's fault when she is raped.
15. T/F Rapists cannot control their sexual urges, "they have to do it."
16. T/F A wife cannot be raped by her husband.
17. T/F Rapes really don't hurt anyone either physically or emotionally.
18. T/F Rape is synonymous with sex.
19. T/F Rape rarely occurs in the United States.
20. T/F All rape victims are female.
21. T/F It really is a woman's fault for not protecting herself from the rape.
22. T/F A woman is safe from rape if she is with a friend or acquaintance.
23. T/F There is no relationship between rape and drug or alcohol abuse.
24. T/F It's O.K. when rape occurs if the man spent money on the woman.
25. T/F It's O.K. when rape occurs if she lets him touch her private areas.
26. T/F It's O.K. when rape occurs if she is hitchhiking.
27. T/F It's O.K. when rape occurs if the rapist is "so turned on that he can't stop."
28. T/F It's O.K. when rape occurs if she has had sex with him before.
29. T/F It's O.K. when rape occurs if they have known each other for a long time.
30. T/F Some women "deserve" to be raped.
31. T/F Rape does very little harm to the victim.
32. T/F Rape is a crime that has been dramatically decreasing in the United States.
33. T/F Based on current statistics, chances are that most U.S. women will never be raped.

34. **T/F** Rapes rarely occur in the victim's or perpetrator's home.
35. **T/F** Very few rapes are planned in advance.
36. **T/F** "Good" "Nice" boys/men don't rape.
37. **T/F** When a woman finally submits to the attack, she is basically giving her consent.
38. **T/F** A convicted rapist usually has no prior history of sexual violence.
39. **T/F** Most rapists can remember exactly what their victim was wearing when they attacked her.
40. **T/F** Young girls and very old women are never raped.

Key

1–40 = false

How well did you do? Grade your quiz and give yourself 1 point for each correct answer. The higher you score (**40** is max) the more accurate your understanding of rape and rape issues tends to be. Go back and carefully consider why you missed the ones you missed. Do the findings from this assessment accurately reflect you, your expectations, and your life experience? Why?

Divorce Risks Assessment Questionnaire

Y ou can do this assessment if you are married, divorced, widowed, dating, engaged, or otherwise committed. Simply put a check mark beside all of the risk factors below that apply to you or your significant other in your current or recent relationship.

_____ 1. I/We don't communicate feelings/needs and wants clearly.

_____ 2. I/We don't resolve differences in a timely and fair manner.

_____ 3. I/We married as teenagers.

_____ 4. I/We got married because of a pregnancy.

_____ 5. I/We lived together (cohabited) before marriage.

_____ 6. I/We have little in common.

_____ 7. I/We have little education beyond high school.

_____ 8. I/We have a low-paying job.

_____ 9. I/We don't spend a lot of time with each other.

_____ 10. I/We married someone of another religion.

_____ 11. I/We married someone of another race or ethnic group.

_____ 12. I/We had a very short courtship (less than three months).

_____ 13. I/We had a very long courtship (more than three years).

_____ 14. I/We have been married before.

_____ 15. I/We have a problem with alcohol/drugs.

_____ 16. I /We have a seriously ill child.

_____ 17. I/We don't care much for religious involvement.

_____ 18. I/We think about getting a divorce often.

_____ 19. I/We talk about getting a divorce often.

_____ 20. I/We have been living apart because of work or other circumstances.

_____ 21. I/We have had problems with extramarital sex (adultery or affairs).

_____ 22. I/We have struggled with unemployment.

_____ 23. I/We have significant differences in our education levels.

_____ 24. I/We have parents who divorced.

_____ 25. I/We have or want no children.

_____ 26. I/We have or want more than five children.

_____ 27. I/We have been married less than five years.

_____ 28. I/We live in or near a city.

_____ 29. I/We have other partners available.

_____ 30. I/We have no sons.

_____ 31. I/We have recently been in combat.

_____ 32. I/We have lost the feeling of love.

_____ 33. I/We have significant financial problems.

_____ 34. I/We have a great deal of consumer debt.

_____ 35. I/We have serious problems with in-laws.

_____ 36. I/We have problems with cruelty, physical, or emotional abuse.

_____ 37. I/We have problems with compatibility in sexual matters.

_____ 38. I/We don't like my partner/each other.

_____ 39. I/We spend more time with friends than each other.

_____ 40. I/We feel that divorce is an appropriate way to solve marital problems.

_____ 41. I/We had a troubled childhood.

_____ 42. I/We married to get away from our family of origin.

_____ 43. We eloped when we married.

_____ 44. I/We met and married shortly after a major life crisis.

_____ 45. I/We depend on family or friends for financial, physical, or emotional support.

_____ 46. I/We have poor relationships with siblings.

_____ 47. I/We live a considerable distance from family of origin.

_____ 48. I/We live considerably close to family of origin.

_____ 49. I/We often look at others as a potential future spouse.

_____ 50. I/We have lots of friends who are divorced.

_____ 51. I/We have serious disagreements on how to raise the children.

_____ 52. We have grown apart.

_____ 53. I/We have different interests and values now.

_____ 54. I/We spend too much time at work.

_____ 55. I/We have gone to friends or family seeking information on how to divorce.

Making Sense of the Risk Factors

Keep in mind that these are correlates of divorce and not causes. If they are present in your life experience, that does not imply that your marriage is doomed. It does imply that you may have to exert more energy in order to deal with them and to avoid their potentially destructive influence in your marriage. Research into why these factors are correlated with divorce will help you to better understand them and how they effect your marriage. Consider each factor you placed a check mark beside using the following questions.

How does this risk factor apply to my marriage?

Is this factor something I can have control over?

If I can have some control over this factor, what specifically am I willing to do about it (i.e.: counseling, therapy, change lifestyle, etc. . . .)?

If I really cannot control this factor, how am I going to adjust my marriage relationship in order to cope successfully with it?

Select which factor you will focus on first by assessing its degree of importance to you in terms of marital success. Start by writing the number of each factor in the boxes below:

Very Critical	Somewhat Critical	Not Sure	Not Very Critical

Now that you have divided your risk factors by their relative importance, choose the most important factor in the "Very Critical" category and create a plan for how you and your spouse will exert your personal energies to control or cope with it. Knowing which risk to look out for is the first step in taking charge of your marriage's destiny.

How Healthy Is My Remarriage/Step Family?

Answer the true/false questions below as accurately as possible as they relate to your step family. This quiz can be taken by married, remarried, or single adults.

1. T/F Relationships with ex-spouses are basically nonconflictual and compatible.
2. T/F Custody issues have been resolved legally.
3. T/F Custody issues have been resolved interpersonally.
4. T/F Financial arrangements have been finalized.
5. T/F Payments are made where they are due (i.e.: child support, alimony, etc. . . .).
6. T/F There are enough economic resources to meet the family's basic needs.
7. T/F Mom and Dad work together on discipline issues for the children.
8. T/F Work in the home is distributed among all family members.
9. T/F Family members can communicate openly about issues as they arise.
10. T/F Realistic expectations exist as far as developing feelings of love for step-family members.
11. T/F There is more discipline in the home than punishment.
12. T/F Each family member is valued and respected in the home.
13. T/F Children are supported in the relationship between them and *both* of their biological parents.
14. T/F When issues of anger and hurt arise they can be dealt with by family members.
15. T/F Mom and Dad often take time away from home to nurture their relationship.
16. T/F Grandparents are accepting of all the children in the family.
17. T/F When Mom and Dad disagree, they do so privately as a couple.
18. T/F Everyone in the home is committed to the success of the family.
19. T/F A healthy system of integrating all monies into the family budget has been established.
20. T/F It is O.K. for children to be loyal to *both* parents.
21. T/F Family members realize that we cannot create a replica of the biological family.
22. T/F Regular family meetings are held to facilitate concerns, problems, scheduling, and so forth.
23. T/F Parents and children have established comfortable titles for each other.
24. T/F Agreements have been made for holidays, birthdays, and other special occasions.
25. T/F We have fun together in the family.
26. T/F We tend to be very flexible in terms of family roles.
27. T/F Each family member has his or her needs for privacy and space met.
28. T/F We have very few rules and flexibility exists when called for.
29. T/F We allow plenty of time for family members to adjust and deal with issues.
30. T/F The relationships between children and all of their grandparents are supported.

Key

Give yourself **1 point** for each True answer. The higher the score the higher the level of function in the step family. Consider all of the questions that were answered False. Which are the most significant to you and other family members who take this quiz? Discuss them together in the context of looking for a healthy strategy for improving upon them. Remember that most step families portrayed on television have scripts already prepared for them to follow. In real life we must choose and act for ourselves if things are to get better.

Do the findings from this assessment accurately reflect you, your expectations, and your life experience? Why?

Who Defines Families and Policies Relating to Families?

Which way does the pendulum swing today?

Please list the current overall governmental and religious stances on each of the issues below.

Government   **Religion**

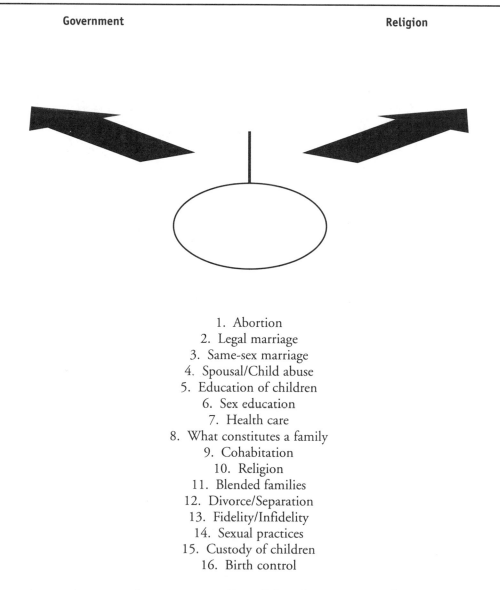

1. Abortion
2. Legal marriage
3. Same-sex marriage
4. Spousal/Child abuse
5. Education of children
6. Sex education
7. Health care
8. What constitutes a family
9. Cohabitation
10. Religion
11. Blended families
12. Divorce/Separation
13. Fidelity/Infidelity
14. Sexual practices
15. Custody of children
16. Birth control

Compare your values to the stance of government and/or religion. Are you more influenced by one, the other, or both?

Challenges of Financial Status

All families, no matter what their financial status, face challenges. One challenge to the wealthy upper-class and the upper-middle-class children is that they often receive everything they want even before they know they want it. This may cause them to lose their hopes and dreams. They may not establish goals for the future and may look toward getting more and more things and are therefore not happy with what they have. They may live for the moment and equate love with acquiring things.

The middle-class and lower-middle-class children may become complacent and not look to improve their life situation. Often both parents may need to work and their children may be raising themselves or are in the workforce themselves.

The poor are discouraged and may not see any way out of their dilemma. Delinquency and crime become a common problem for adolescence. They do not have good health care, legal counsel, or extended support systems. They may still be happy and not worry about problems related to affluence.

Choose one of the economic levels and write what you would do to help your children have a successful, happy, and productive childhood at that level.

Financial Status _____

Goals for this level:

1.

2.

3.

4.

Precautions:

1.

2.

3.

Positive influences:

1.

2.

Your Projection for the Family in 2030

I n 1960, a family sociology class was asked how they would predict the makeup of the family in the year 2000. The class felt the family would no longer be a unit as they knew it. It was surmised that a group of people would live together with free love, children would be raised in communal care, and all would be the extended family. This thinking was based on the hippie movement of the 1960s. Some of that prediction has come true with single parents and daycare.

How do I predict the United States average family to be in 2030	
Role of Women	Role of Men
Role of Children	Role of Extended Family

How would I like my family to be in 2030	
Role of Women	Role of Men
Role of Children	Role of Extended Family

Write a brief paragraph describing your family in 2030. Include its size, family members, location, living arrangements, financial status, and a typical daily routine.

Assessment Accomplishing Record

Place a check mark beside the assessments and assignments you completed:

_____ Genogram

_____ Gender Talk Patterns

_____ Gender Talk Patterns—Partner, Family Member, Friend Version

_____ Gender/Androgyny Role Attitude Assessment

_____ Personal Men's Issues Assessment

_____ Softening Patriarchal or Matriarchal Heads of Families

_____ Television Messages About Gender Roles

_____ Planning Your Family Life Cycle

_____ Life Cycle

_____ Nontraditional Life Cycles: Variations on the Theme

_____ Rites of Passage

_____ Using the SMART Approach to Evaluating Information

_____ Your Dispositions Toward Theories

_____ Critical Life Events Log

_____ How Others Influence Your Self-Concept

_____ How Colonial Is Your Family?

_____ Food and Meals in My Home

_____ Family Foods and Meal Traditions

_____ Attitudes About Food in Your Family of Origin

_____ How Metropolitan Are You?

_____ Singles Well-Being Assessment

_____ How Safe Was the Home You Grew Up In?

_____ Family of Origin Enmeshment

_____ Your Religiosity Continuum

_____ Your Partner's Religiosity Continuum

_____ Sexual Harassment Awareness Quiz

_____ Quality of Health Care Coverage

_____ The Value of Education in My Family: Self-Version

_____ The Value of Education in My Family: Partner Version

_____ Net Worth Data Sheet

_____ Your Intergenerational Mobility

_____ Your Spouse's Intergenerational Mobility

_____ Your Intragenerational Mobility

_____ Your Spouse's Intragenerational Mobility

_____ My Social and Financial Position

_____ Who Are the Poor in America?

_____ Social Reproduction

_____ The Self-Portrait: An Exercise in Appreciating Our Self and Our Groups

_____ The Super Self-Portrait

_____ Understanding the Self- and Super-Self Portrait

_____ My Predispositions to Prejudice: Self-Version

_____ My Predispositions to Prejudice: Partner Version

_____ Understanding Problems at the Group Level: An Exercise in Overcoming Prejudice

_____ Your Family Race, Ethnic, and Cultural Heritage

_____ Your Mother's, Father's, and Your Own Ancestral/Biological Heritage

164

_____ My Mother's Ancestral/Biological Heritage

_____ My Father's Ancestral/Biological Heritage

_____ How Did You Meet Each Other?

_____ Intimacy Assessment Questionnaire

_____ Quality Communications Assessment

_____ How Are You Attracted to Partners?

_____ Speaking and Hearing Your Love Styles: Self-Version

_____ Speaking and Hearing Your Love Styles: Partner Version

_____ What Type of Lover Are You?

_____ Understanding the Love Type Categories and Their Meanings

_____ Ideal versus Practical Love Styles

_____ How Healthy Are Your Break Ups?

_____ Open and Closed Families

_____ Boundary Maintenance Questionnaire

_____ My House, My Boundaries

_____ Boundary Maintenance Worksheet

_____ What Do You Look for in Your Ideal Mate?

_____ My Ideal Wife

_____ My Ideal Husband

_____ Homogamy Self-Assessment

_____ Homogamy Partner Assessment

_____ Processing the Homogamy Assessments

_____ Quality of My Dating Experiences

_____ Fiancé/Fiancée Communication Awareness Checklist

_____ Marital Anxiety Attitudes Assessment

_____ Where You Learned About Sex and Reproduction

_____ Television Messages About Sexuality

_____ Childbirth Preferences Questionnaire

_____ Death and Grief Assessment

_____ How Strong Is This Relationship?

_____ Marital Expectations Assessment

_____ Do I Have a Strong Family?

_____ An Ideal Family Holiday and Food

_____ Recreation Assessment

_____ Who Has Possession of the Remote Control? Self-Version

_____ Who Has Possession of the Remote Control? Partner Version

_____ My Ideal Obituary/Life Goals

_____ How Has Parenthood Changed You?

_____ Single Parenting Assessment

_____ Fatherhood Patterns Assessment

_____ Motherhood Patterns Assessment

_____ Processing the Fatherhood and Motherhood Patterns Assessments

_____ Children and Attachment

_____ Assessing the Use of Blame in Your Relationships

_____ Marital Violence Risks Assessment

_____ Rape Awareness Quiz

_____ Divorce Risks Assessment Questionnaire

_____ How Healthy Is My Remarriage/Step Family?

_____ Who Defines Families and Policies Relating to Families?

_____ Challenges of Financial Status

_____ Your Projection for the Family in 2030